MATHEMATICS RESEARCH DEVELOPMENTS

CLUSTER EFFECTS IN MINING COMPLEX DATA

MATHEMATICS RESEARCH DEVELOPMENTS

Additional books in this series can be found on Nova's website
under the Series tab.

Additional E-books in this series can be found on Nova's website
under the E-books tab.

COMPUTER SCIENCE, TECHNOLOGY AND APPLICATIONS

Additional books in this series can be found on Nova's website
under the Series tab.

Additional E-books in this series can be found on Nova's website
under the E-books tab.

MATHEMATICS RESEARCH DEVELOPMENTS

CLUSTER EFFECTS IN MINING COMPLEX DATA

M. ISHAQ BHATTI

Nova Science Publishers, Inc.

New York

NOTICE TO THE READER

The Publisher has taken reasonable care in the preparation of this book, but makes no expressed or implied warranty of any kind and assumes no responsibility for any errors or omissions. No liability is assumed for incidental or consequential damages in connection with or arising out of information contained in this book. The Publisher shall not be liable for any special, consequential, or exemplary damages resulting, in whole or in part, from the readers' use of, or reliance upon, this material. Any parts of this book based on government reports are so indicated and copyright is claimed for those parts to the extent applicable to compilations of such works.

Independent verification should be sought for any data, advice or recommendations contained in this book. In addition, no responsibility is assumed by the publisher for any injury and/or damage to persons or property arising from any methods, products, instructions, ideas or otherwise contained in this publication.

This publication is designed to provide accurate and authoritative information with regard to the subject matter covered herein. It is sold with the clear understanding that the Publisher is not engaged in rendering legal or any other professional services. If legal or any other expert assistance is required, the services of a competent person should be sought. FROM A DECLARATION OF PARTICIPANTS JOINTLY ADOPTED BY A COMMITTEE OF THE AMERICAN BAR ASSOCIATION AND A COMMITTEE OF PUBLISHERS.

Additional color graphics may be available in the e-book version of this book.

Library of Congress Cataloging-in-Publication Data

Bhati, M. Ishaq.
 Cluster effects in mining complex data / M. Ishaq Bhatti.
    p. cm.
 Includes bibliographical references and index.
 ISBN 978-1-61324-482-1 (hardcover)
 1. Cluster analysis. 2. Econometrics. 3. Data mining. I. Title.
 QA278.B52 2011
 519.5'3--dc23
                          2011012590

Published by Nova Science Publishers, Inc. †New York

CONTENTS

LIST OF TABLES

PREFACE

In the last few years there has been a significant growth of research in mining complex databases, which has been generated from pattern studies, longitudinal, time series, panel and/or sample surveys in their application in health, medical and other social sciences. Often the data arising from these areas are naturally in blocks or clusters. When regression models are applied to data obtained from such clusters, the regression errors can be expected to be correlated within ultimate clusters and subclusters. As one would expect, ignoring such correlation can result in inefficient estimators, and in seriously misleading confidence intervals and hypothesis tests. This book attempt to develop a unified model and efficient estimation and testing procedures for the cluster effects associated with multi-factor clustered data. The usefulness of the method is being demonstrated by applying it on the real data set to assess the performance of the simulation methods and testing techniques.

This book considers several aspects of hypothesis testing issues associated with the cluster effects using the multi-stage linear regression model. The statistical universe in these fields, as indeed in many others, whilst conceived as whole is often comprehended as being constituted by a number of natural clusters. For example, the population of a country falls naturally into the clusters of states, cities, districts, subdistricts, or urban and rural areas. Similarly, industries are often perceived as a complex of clusters and subclusters, with each concerned with the well defined processes of manufacturing and production process.

When regression analysis is used to analyse data obtained from such clusters, subclusters or groups, the residuals are often found to be correlated within ultimate clusters. This correlation is sometimes constant across clusters, with its coefficient referred as the intracluster or equicorrelation coefficient. This correlation in the error term, and is also called the cluster effect. The main aim of this book is to develop and investigate testing and estimation procedures for such cluster effects.

This book is divided into three parts. Part one of the book reviews literature and recommends the use of a model-based approach in the analysis of blocked or clustered survey data, suggesting a unified regression model, which could be equally useful for survey statisticians and at times panel-data econometricians. It summarises the theory of point optimal testing and suggests how such optimal tests can be applied to the unified regression model in the subsequent chapters. In chapters three and four the small-sample properties of optimal tests for cluster effects in Standard Symmetric Multivariate Cluster (SSMC) and Symmetric Multivariate Cluster (SMC) are developed and investigated respectively. The SMC distribution is a multivariate distribution, in which all components of a k dimensional random vector, y, and have equal means, equal variances and all covariances between components which take the same value. These common covariances give rise to a common equicorrelation coefficient, ρ which is called the cluster effect. In this second part of the book, some optimal tests for two- and three-stage SSMC and SMC distributions for detecting intracluster and intrasubcluster are developed and investigated.

Part two of the book extends SSMC model to linear cluster regression models, in which the disturbances follow two-, three- and higher-stage clustered distributions are explored. An attempt is made to find an optimal exact test for the intracluster correlation for the two stage linear regression (2-SLR) model, which is then generalised to a multistage linear regression model by using Point Optimal Invariant (POI) and the Locally Most Mean Powerful Invariant (LMMPI) tests approaches.

Part three of the book consists of three chapters. Chapter 7 considers LNSG model and various associated tests discussed in earlier chapters. Chapter 8 concentrate on random coefficient stochastic model for cluster effects while chapter 9 addresses the problem for testing for change point. The final chapter contains some concluding remarks.

Melbourne
14 July, 2011

                                              M. ISHAQ BHATTI

1. INTRODUCTION

1.1. PREAMBLE[1]

Sample surveys are now being routinely used as a tool by economists, business consultants and statisticians to gather data and interpret it before making important decisions, or giving much needed advice. Due to rapid growth in computing power, the statistical analysis of such data is becoming more sophisticated. Often, among other methods, linear regression methods are used to analyse such bodies of data. This is frequently done without due regard being given to the complications that can arise. For example, such data sets may be subject to heterogeneity, as it is common for data for economic research to be collected in clusters. In such a case, data elements from different clusters may be affected by any number of diverse effects, such as climate, soil, prices, etc. The similarities within a cluster can give rise to what is known as intracluster correlations or 'cluster effects' in the error terms of regression models. If such effects are strong, the standard regression techniques are in appropriate.

The main aim of this book is to develop and investigate diagnostic tests for such cluster effects. Just as a doctor runs diagnostic tests on a patient to see if there is a specific medical problem, an econometrician/statistician runs tests on the regression model to see whether an intracluster correlation is present. The simplest form of an intracluster correlation arises from two-stage cluster sampling, and with three and multi-stage clustering as well. In a national

---

[1] Material of this book has originated from various published work (out of print book, Bhatti, (1995), book chapter, Bhatti (2004), and joint article (Bhatti and Wang, 2005), among others).

survey the largest clusters might be, for example states, and followed by regions, postal districts and finally the streets. A national survey is an example of a four-stage cluster sampling framework. As a result of state laws and practices, one might also expect an intracluster correlation within states. Diagnostic tests for an intracluster correlation within smaller clusters such as e.g., regions and/or postal districts have been developed and are evaluated using computer simulation techniques.

Generally, it is common for researchers to ignore the existence of such cluster effects, arguing perhaps that the amount and impact of these effects are very small and thus unlikely to affect the analysis significantly. As pointed out by Holt and Scott (1981), followed by Deaton and Irish (1983), Moulton (1990) and Bhatti (1994, 2001) this small residual cluster correlation can cause standard errors of the Ordinary Least Squares (OLS) estimator to be seriously biased downwards. These can result in spurious findings of statistical inference, and in turn lead to misleading conclusions. This book investigates the small sample power properties of some diagnostic tests for detecting the existence of such cluster effects from the multistage linear regression model and proposes some extension.

The theory of statistical hypothesis testing enables us to develop such diagnostic tests. No matter whether these tests are classical, namely the Wald Likelihood Ratio (LR) and Lagrange Multiplier (LM), or the Cox's Non-nested tests (Cox, 1961 and 1962), Hausman's Specification tests (Hausman, 1978), White's Information Matrix tests (White, 1982), Conditional Moment tests (Newey, 1985 and Tauchen, 1985) and Point Optimal tests (King, 1987), our ultimate concern is with the test's power. This is because the power of a test determines its accuracy and the chance of making a correct decision by incorporating knowledge of the signs of the parameters under the test.

Among the classical tests, the LM test is the most popular mainly due to its ease of calculation, but its size and power can be distorted in smaller samples. Therefore, our alternative choice is to construct a test, which ensures certain known optimal power properties. Examples of such tests are already available, and include Locally Best (LB) tests (Neyman and Pearson, 1936 and King and Hillier, 1985), the Point Optimal (PO) tests (King, 1987) and Locally Most Mean Powerful (LMMP) tests (King and Wu, 1990). These tests are reviewed in detail in the next chapter, but in light of their names (and claims), it is important to note that each of these tests has a small sample optimal power property. Ideally, wherever the testing strategy is undertaken and small sample optimal tests are conducted, it is worth comparing the power properties obtained with those obtained by large sample tests (e.g. LM). At a practical level, the actual size and power properties yielded by small sample optimal tests can suggest whether a larger sample test is warranted.

Deaton and Irish (1983) in their study considered a linear regression model based on a two-stage clustered sampling design, and suggested the use of the one-sided LM (LM1) test, whilst King and Evans (1986) in their study showed that the test is LB invariant (LBI). SenGupta (1987) considered a model based on standard symmetric multivariate cluster(SSMC) distribution and proposed a LB test for testing positive values of the equicorrelation coefficient for this model. Based on these studies the following issues were found worthy of further consideration.

1. How can PO tests, based on SSMC and SMC distribution models for the same testing problem considered by SenGupta, be developed? How can LMMP tests be constructed when the number of parameters being tested is more than one?

2. Is it possible to construct PO Invariant (POI) tests for the same testing problem considered by King and Evans, and if so, how well do they perform relative to other existing tests? Can the LMMP invariant (LMMPI) tests be constructed for testing multi-parameter extensions of the King and Evans' model?

3. How well does the LMMPI test perform for testing multi-cluster effects whilst dealing with a three-stage model? How well do the POI and the LM tests perform when the number of parameters being tested is more than one, in relation to a three-stage model? How can POI and LMMPI tests be constructed for a multi-stage model?

4. Is it possible to construct optimal tests for testing correlated data for LNSG model which comprised of large numbers of short samples?

5. How to handle stochastic coefficient model when data is equicorrelated/

6. How well nonparapetrics tests for change point problem performed?

This book attempts to answer these questions in the subsequent chapters. It should be noted that the principle of invariance is used throughout this book to eliminate the nuisance parameters, wherever necessary, and in order to reduce the dimensions of the various testing problems under consideration.

1.2. AN OUTLINE OF THE BOOK

In chapter two, the historical background of sample surveys and their associated problems, and a comparison between design-based and model-based approaches are presented. A unified regression model is also developed which shows that the modelling of cluster effects in survey data are similar to that of individual effects in panel data. It is noticed that with a minor difference in concept and a change of notation, both models can easily be used as substitutes for each other. In the search of Uniformly Most Powerful (UMP) tests, this chapter also reviews some of the existing tests and discusses the new optimal tests, which make use of computer technology to maximize and/or improve their power under the alternative parameter space.

In chapter three the problem of testing for non-zero values of the equicorrelation coefficient of SSMC distributions is considered. The LB, Beta Optimal (BO) and LMMP tests for the case of two-stage SSMC distribution model will also be constructed.

Next the power of the two versions of the BO test with that of SenGupta's (1987) LB and the power envelope (PE) is compared, and it thereby will demonstrate the superiority of the BO tests. This chapter extends the two-stage SSMC model to a three-stage model, when more than one parameter is being tested, and constructs PO and LMMP tests. In addition, this chapter also discusses more complicated matters, such as testing subcluster effects in the presence of main cluster effects, in the case of the three-stage SSMC model.

In chapter four analogous tests (i.e. POI and LMMPI tests) for the two-and three-stage SMC distributions will be developed. The findings in this chapter suggest that the POI test for the case of the two-stage model is UMP invariant (UMPI). The most attractive features of both the POI and the LMMPI tests are that their power and critical values are obtainable from the standard F distribution. For the three stage SMC model a POI test for testing subcluster effects in the presence of main cluster effects is developed. The subsequent two chapters investigate various aspects of statistical inference in the linear regression model with disturbances that follow SMC distributions.

In chapter five, the size and small sample power properties of the numerous statistical tests associated with the two-stage linear regression model for detecting simple intracluster correlation or equicorrelation are investigated. The results suggest that the POI test is marginally better than the LM1 test for small and moderate sample sizes. Both the LM1 and the POI tests are approximately UMPI for some selected cluster sizes. It is also noted that the LMMPI test for testing the hypothesis for zero equicorrelation against the alternative that the correlation may differ from cluster to cluster is equivalent to that of the LM1 test. An empirical power comparison of the LM1, two-sided LM (LM2), Durbin-Watson (d), modified d (d*), POI tests and the PE show the relative strengths of these tests.

Chapter six extends upon the model of chapter five to cover a more general situation, and constructs POI and LMMPI tests for zero equicorrelation over different cluster (ρ_1) and over different subcluster (ρ_2). This chapter also discusses how close the power of the LMMPI test is to that of the PE. The problem of testing subcluster effects, in the presence of main cluster effects is also explored in this chapter, together with the small sample power properties of different statistical tests associated with the three-stage linear regression model. Monte Carlo studies are used to investigate the power performances in small samples, with this chapter also generalizing these tests for the case of multi-stage linear regression models and developing some optimal tests.

Chapter seven generalizes Cox and Solomon's (1988) model and suggests optimal tests for testing serial correlation coefficient in large numbers of small samples. Chapter eight extend fixed effect model by incorporating stochastic component in the regressors and develops random coefficient models capturing cluster effects, and suggests efficient estimation procedures to be applied in real life data. Chapter nine concentrates on the issue of change point problem when data arises from SSMC and SMC models. Some selected nonparametric tests are developed and empirical power comparisons of these tests are conducted to measure the strength of each test under consideration. The final chapter provides some concluding remarks.

2. MODELLING CLUSTERED SURVEY DATA

2.2. INTRODUCTION

In recent years, there has been an increasing use of regression analysis on survey data in empirical econometrics, statistics, market research, social, physical and biological sciences.

Often the data arising from these areas are naturally in clusters. When regression is applied to data obtained from such clusters, sub-clusters or multi-clusters (i.e. from multi-stage sample design), regression errors can be expected to correlate within ultimate clusters or sub-clusters. As one would expect, ignoring such cluster correlations can result in an inefficient estimator, and seriously mislead confidence intervals and hypothesis tests. Part of the growing literature in this area is based on traditional sampling frameworks such as, for example, studies undertaken by Konijn (1962), and Kish and Frankel (1974). Others such as Scott and Holt (1982), Deaton and Irish (1983), King and Evans (1986), Christensen (1986, 1987a, 1987b), Moulton (1986, 1990) and Dorfman (1993), Rao et. al. (1993) Wu and Bhatti (1994), Bhatti (2000, 2001, 2004) whereas focus on a model based approach.

This book follows a model-based approach in order to deal with regression analysis problems of sample survey data. The literature survey of the model based approach work reveals that, most of the above mentioned authors have considered only the simplest form of intracluster correlation arising from the two-stage cluster sampling. In subsequent chapters, an extension of their work in two directions is explored. Initially, a PO test for testing intracluster correlation in a two-stage model, to achieve comparability with the other existing tests, is developed. Following this, and having found such testing procedures superior in terms of power performance, a PO test is used on a three-stage model to check its viability and practicality.

In reality, sometimes two-stage models are woefully inadequate. For example, in an investigation of a national population with a basic unit of household, a two-stage model would be well neigh useless. What is needed in such as situation is a model that allows for clustering and subclustering at regions, states, wards, postal districts, streets or similar levels. In other words, a multi-stage model is needed. Hence, this book considers the use of a three-stage model in order to demonstrate further viability of our testing procedures, and to understand the consequences of such an extension. Furthermore, the use of a general multi-stage model is explored.

There are three main objectives of this chapter, which are to firstly develop a unified linear regression model, secondly to review the theory of optimal hypothesis testing and finally to survey several existing diagnostic tests for detecting cluster effects associated with our model. The chapter starts with an account of the history of sample-surveys and their use in regression analysis. The choice of using a model-based approach when working with survey data is justified as well in the process, with problems associated with clustered data also raised. A unified regression model is then developed, which can be used by statisticians dealing with survey data, and in special cases can also be used by econometricians dealing with time series cross sectional and/or panel data. This section of the chapter will also discuss the variety of models involving fixed and random coefficients models with one and two-way error components models, and finally the multi-stage linear regression model. The theories of optimal testing and discussion on some problems associated with the theory of PO testing is also summarized in this chapter, with the latter part of the section reviewing some articles which construct exact tests for cluster effects from SSMC, SMC and two-stage linear regression models. In the final section of this chapter concluding remarks are made.

2.3. HISTORICAL BACKGROUND OF SAMPLE SURVEYS

The idea of sample surveying was developed by Kiaer (1895) and was adopted by the International Statistical Institute in its first meeting in Berlin in 1903. Bowley (1906) followed this development, and supplied the theory of inference for survey samples to Neyman (1934) who was the first to be concerned with the choice between alternative sampling procedures for survey data, such as stratified, purposive and cluster sampling. Others who contributed significantly to sample surveys were Cochran (1942) who dealt with ratio and regression estimators; Hansen and Hurwitz (1949) who developed the selection procedures with probability proportion to size; Madow and Madow (1944) and Cochran (1946) who considered the method of systematic selection.

Due to its wide application in different fields of human endeavour, quite a number of textbooks have been devoted to the subject of sample surveys. Among the most notable texts are Yates (1949), who was the first, followed by Deming (1950), Cochran (1953), Hansen, Hurwitz and Madow (1953), Sukhatme (1954) and Kish (1965). The latest editions of these works are largely refinements and improvements on the originals, although they contain some new areas of thought, which have been generated from further research in the area. For example, Godambe (1966), Royall (1968, 1970), Hartley and Rao (1968, 1969), Scott and Smith (1969) and Särndall (1978) have used the idea of finite and superpopulation in relation to the case of multistage and other survey designs.

Very interesting work has also been done in the last few decades on the method of data analysis of complex sample surveys, particularly so in the field of regression analysis the study by Kish and Frankel (1974) and Pfefferman (1985), which represents a new approach. Kish and Frankel and Pfefferman investigated the impact of an intracluster equicorrelation coefficient on the regression analysis by using the standard sample survey framework. They point out in their study that, if the intracluster correlation is ignored, then the estimates of the calculated standard errors of the regression coefficients underestimate the true standard errors.

Through the 1980's the practitioners and theoreticians discussed the controversial issue of whether a design-based or a model-based approach should be adopted for regression analysis of sample survey data. The difference between the model-based linear least squares prediction approach and the design-based sampling distribution approach is discussed in the next section.

Since the 1980s, the area of regression analysis approaches towards sample survey data in model-based and design-based models is more often explored, with Shao and Wang (2002), Davern et al (2007) and Desai and Begg (2008) being some studies which more recently look more in depth at correlation coefficients and cluster effects in such approaches.

2.4. MODEL-BASED VERSUS DESIGN-BASED APPROACHES

It is well known that the design-based approach in regression means a relationship between the dependent variable, y and the independent variable, x, where x is obtained from well defined clusters of a finite population. In this approach, each observation is weighted by the reciprocal of its probability of being included in the sample, in order to make an inference about the finite population on the basis of the observed sample. Here unbiased estimators of the intercept and the slope coefficient are obtained by taking the expectation of all possible samples and a consistent estimator, which can be obtained by using Cochran's (1953) definition.[1] In the design-based approach, it is believed that the allocation of weights in the sample selection procedure make the inference about population parameters inefficient and hence it does not lead to Best Linear Unbiased Estimators (BLUE).

In contrast to the design-based approach, the model-based approach always assumes that there exists a data generating process for the variables and therefore there is no finite population. For example, when money demand, GDP, or any other macroeconomic variable is observed, there is no finite population. This leads to the use of a stochastic model to explain the data generating process as correctly as possible. The simplest stochastic model is

$$y_i = \alpha + \beta x_i + \varepsilon_i$$

where the ε_i's are i.i.d.$(0, \sigma^2)$. If the xi's are exactly known, the OLS estimator is the minimum variance linear unbiased estimator (or BLUE) for β and $\tilde{\alpha}$.

In this model, the word 'unbiased' refers to the expectations of overall possible realizations of the stochastic process, while 'consistent' is used in the usual econometric or statistical sense.[2]

An interesting evaluation of the model-based and design-based approaches is given by Royall and Cumberland (1981), who compares the variance estimates of both approaches, with the empirical results favouring the use of the model-based approach due to its efficiency and superiority. A further advantage of the model-based approach is that it can be applied using existing

---

[1] An estimator is consistent in this approach if the estimate becomes exactly equal to the population value, when n = N, that is, when the sample consists of the whole population.
[2] A consistent estimator is an estimator, which converges in probability as the sample size increases with a decreasing variance to the parameters of which it is an estimator.

software regression packages. Examples of such computer packages are SOLAS and PROC MI (Horton and Lipsitz, 2001), FORTRAN (Dunlap et al, 2004) and R package (Nimon et al, 2008). This book, has adopted the procedure of using the model-based approach for the analysis of clustered survey data.

Problems of Clustered Data

Scott and Holt (1982) considered the effects of a two-stage design on the OLS estimator, particularly on efficiency and on standard errors. They showed that the OLS estimator is consistent but inefficient, with the standard estimator for the variance-covariance matrix consistent only if $\rho = 0$. Kloek (1981), Greenwold (1983) and Moulton (1986) have analysed the magnitude of the bias. Kloek, Greenwold and Moulton's studies show that the magnitude of the downward bias for the standard errors increases with an increase in the average cluster size, the intra-cluster or equicorrelation of disturbances, and the equicorrelation of the regressors.

Despite these problems, survey data would always be preferred by the researchers due to its availability from secondary sources and the time and cost advantages over time-series cross-sectional and/or panel data sets. In part to the high-tech and competitive markets, the analysis of survey data is very important, there is a need for diagnostics tests on the validity of such models is a basic requirement for researchers. Discussion on some existing tests and proposed optimal tests in this area are detailed later in the chapter. Prior to this, the following section will present a discussion on a unified linear regression model, which will be used throughout this book in a different format.

2.5. A UNIFIED[3] REGRESSION MODEL FOR SURVEY DATA

An unified regression model means an equation in which the grouping/clustering of homogeneous characteristics is reflected in the disturbance term, u_{ij}.

---

[3] It unifies regression modeling of survey and panel data, as it can be noted from equation 2.4.

Two-Stage Linear Regression Model

In this model it is assumed that n observations are available from a two-stage sample with m blocks or clusters. Let $m(i)$ be the number of observations from the ith cluster so that $n = \sum_{i=1}^{m} m(i)$. Therefore the simplest model of this form would be

$$y_{ij} = \sum_{k=1}^{p} \beta_k x_{ijk} + u_{ij} \left(i = 1, 2, ..., m, \ j = 1, 2, ..., m(i) \right),   \text{(2.1)}$$

in which i is the cluster identifier, j is the observation identifier in the given cluster, β_k are unknown coefficients, x_{ijk} for $k = 1, 2, ...,$ and p is observations on p independent variables, the first of which is a constant. It is assumed that u_{ij}'s are independent between clusters but equicorrelated within clusters. Hence

$E(u_{ij}) = 0$ and

$$\text{cov}(u_{ij} u_{st}) = \begin{cases} \sigma^2 & (i = s, j = t), \\ \rho\sigma^2 & (i = s, j \neq t), \\ 0 & (i \neq s). \end{cases}   \text{(2.2)}$$

Here ρ in (2.2) is the equicorrelation coefficient of the disturbances.

Models of this form have been considered by Fuller and Battese (1973), Fuller (1975), Campbell (1977), Deaton and Irish (1983), Christensen (1984), King and Evans (1986), Honda (1989) and Bhatti (1994, 2000) among others. A detailed discussion on this model (and its simplest form) is given in chapter five.

If it is assumed that $x_{ijk} = 0$ or $\beta_k = 0$, then the model (2.1) will become SMC distribution model. The further assumption of $\sigma^2 = 1$ in (2.1) through (2.2) make this as a SSMC distribution model (see Bhatti and King, 1990). The definitions and examples of both these distributions are given in the next section.

Standard Symmetric Multivariate Cluster Distribution

The purpose of this subsection is to define SSMC and SMC distributions and then investigate some of their applications in the field of statistics and econometrics. It is important to note that these distributions arise naturally in biometrics, education, genetics, psychology and related areas. Examples include the analysis of missing observations in time series (Sampson, 1976, 1978), generalized canonical variable analysis (SenGupta, 1983) and distributional modelling of repeated failure time measures (Crowder, 1985).

These distributions have been applied in econometrics and statistics literature, for example by SenGupta (1987, 1988), Williams and Yip (1989), Bhatti and King (1990), Bhatti (1995a) and Bhatti et al (2006), with some findings reported in this section from Bhatti et al. To econometricians and statisticians, the best known model based on these distributions is a special case of pooled time series and cross-sectional data, with another important application being a regression analysis of data drawn from two (or higher) stage clustered surveys. For example, when census data is used, a cluster may be defined by a standard city or an irregularly shaped area with identifiable political or geographical boundaries, or it can be a group of industries or occupations etc. For the latter type of data, observations are correlated within ultimate clusters. This correlation is called the intracluster, intraclass, equi-, uniform or familial correlation coefficient, ρ. Ignoring such a correlation can lead to seriously misleading confidence intervals and hypothesis tests based on inefficient OLS estimates, (e.g., see Walsh (1947) and Halperin (1951)), and also produces inefficient forecasts. In the next section these distributions are defined.

Definitions and Examples of SSMC Distribution

The SMC distribution is a multivariate normal distribution in which all the components of a k-dimensional random vector, y, have equal means, equal variances, with all covariances between components take the same value (see Rao, 1973, p. 196). These common covariances give rise to a common correlation coefficient, ρ, which is called an intracluster correlation coefficient. Thus, if $E(y) = \mu$ then SMC model can be written as

$$y \sim N\left(\mu, \sigma^2 \sum (\rho)\right)$$

where[4]

$$\sum(\rho) = \begin{pmatrix} 1 & \rho & \rho & .... & \rho \\ \rho & 1 & \rho & & . \\ \rho & \rho & 1 & & . \\ . & & . & & . \\ . & & & . & . \\ \rho & \rho & \rho & ...... & 1 \end{pmatrix} \tag{2.3}$$

Similarly, a k-dimensional random vector y can be said to follow a SSMC distribution if it follows a SMC distribution, and its components' means and variances are at zero and in unity, respectively, i.e., $E(y) = 0$ and $\sigma^2 = 1$ and hence SSMC distribution model can be expressed as

$$y \sim N\left(0, \sum(\rho)\right)$$

where $\sum(\rho)$ is given by (2.3). Though the literature on SMC distributions is quite extensive, no test for ρ has been proposed for SSMC distributions except the SenGupta's (1987) LB test and the Bhatti and King's (1990) beta optional test. A discussion on these tests is given later in this chapter.

Sampson (1976, 1978) has considered theoretical applications of SSMC distributions and developed the simple best asymptotic normal (BAN) estimation procedure for autoregressive, moving average and intraclass or equicorrelated models. Sampson notes that SSMC distributions arise naturally from multivariate models in which means and variances of individual variables are known, thus allowing these variables to be standardized. Such standardizations are always made, and play an important role in the techniques for reduction of dimensionality, e.g., in canonical variables (Anderson, 1984) and generalized canonical variables analysis (SenGupta, 1981, 1983).

This distribution is of interest to both theoreticians and practitioners for several reasons. The practical application of the SSMC distribution occurs when there are many observations of the individual variables but, because of

---

[4] The structure of the variance-covariance matrix of the observed vector, y, is identical to that of the residuals in panel and survey data, Balestra and Nerlove (1966) and Bhatti (1994), respectively.

historical, financial or practical reasons, there are comparatively too few sets of joint observations. These individual observations can be used to obtain excellent estimates of means and variances, which allow one to proceed as if these estimates are the true values. Such practical examples can be found in time series analysis, analysis of missing observations, psychometrics, generalized canonical variables, and in biometrics. For related results on models which follow this distribution, one may refer to Wilks (1946), Sirivastva (1965), and SenGupta (1987, 1988). Another example of the use of such a distribution can be found in Crowder (1985) who gives a distributional model for repeated failure time measurements. Further, the SSMC distribution provides a practical example of a curved exponential family[5], and illustrates some associated difficulties and techniques concerned with inference, particularly when dealing with the testing of hypotheses.

One-Way Error Component Model

The model (2.1), under (2.2) is similar to that of the random effects model or one-way error component model used by econometricians in the analysis of panel data.

A simple (re)formulation appropriate in this case is

$$y_{it} = \sum_{k=1}^{p} \beta_k x_{itk} + u_{it}, \, (i = 1, 2, ..., N, t = 1, 2, ..., T)$$

where,

$$u_{it} = \mu_i + v_{it},$$ (2.4)

in which $i = 1, 2, ..., N$, is where N stands for the number of individuals (e.g. households) in the sample, and $t = 1, 2, ..., T$, where T stands for the length of the observed time series. Each of the μ_i's, $(i = 1, ..., N)$ are called an individual

---

[5] Efron (1975) considered arbitrary one-parametric families and attempted to quantify how nearly 'exponential' they are. Since it is well known in the literature that one-parametric exponential families have very nice properties for estimation, testing and other inference problems, statistical curvature is identically zero for exponential families and positive for non-exponential families. Statistical curvature is closely related to Fisher (1925) and Rao's (1962, 1963) theory of second-order efficiency.

effect and v_{it} is the usual (white noise) error term. In this reformulation (at this stage) it is assumed that every cluster has the same number of observations (T).

The basic outline of the model (2.4) has been drafted by its pioneers Balestra and Nerlove (1966), Wallace and Hussain (1969) and Maddala (1971). They assume that:

1 The random variables μ_i and v_{it} are mutually independent.

2 $E(u_{it}) = 0$. This implies that $E(\mu_i) = 0$ and $E(v_{it}) = 0$

3 $\mathrm{var}\,(\mu_i) = \begin{cases} \sigma_\mu^2, \text{for } i = i' \\ 0, \text{otherwise} \end{cases}$

4 $\mathrm{var}\,(v_{it}) = \begin{cases} \sigma_v^2, \text{for } i = i', t = t' \\ 0, \text{otherwise.} \end{cases}$

In comparing (2.1) with (2.4), it is noted that $u_{it} = \mu_i + v_{it}$,

$$\sigma^2 = \sigma_\mu^2 + \sigma_v^2, \text{and } \rho = \sigma_\mu^2 / \sigma^2$$

The only difference between (2.1) and (2.4) is that in model (2.4) the ith cluster consists of the time series of the ith individual and the number of observations in a 'cluster' is T, the length of the time series. In econometrics literature, this model is also called the one-way error component model. Useful references on this include Hsiao (1986), Honda (1989) Moulton and Randolph (1989), Baltagi and Li (1991) Körösi, Mátyás and Székely (1992), with this model frequently used to model panel data in econometrics literature. In chapter five this model is discussed in detail, and is referred to as the 2SLR model.

Three-Stage Linear Regression Model

This subsection extends the 2SLR model of the previous subsection to more general situations, where each cluster is divided into subclusters and one can use the error component to capture these cluster/subcluster effects to form a three-stage cluster sampling design model. The three-stage linear regression (3SLR) model is expressed as

$$y_{ijk} = \sum_{\ell=1}^{p} \beta_\ell x_{ijk\ell} + u_{ijk} \ (i=1,...,m, \ j=1,...,m(i), k=1,...,m(i,j)) \qquad (2.5)$$

in which i is the cluster identifier, j is the subcluster identifier in the ith cluster, k is the observation identifier in the jth subcluster of the ith cluster such that

$$n = \sum_{i=1}^{m} \sum_{j=1}^{m(i)} m(i,j),$$

where β_ℓ are unknown coefficients and x_{ijkl} for $l=1,...,p$ are observations on p independent variables, the first of which is a constant. It is assumed that u_{ijk} is normally distributed with a mean of zero, and its variance-covariance structure is given by

$$\text{cov}(u_{ijk} \, u_{rst}) = \begin{cases} 0, \text{(for } i \neq r, \text{ and } j,s,k \text{ and } t) \\ \rho_1 \sigma^2, \text{(for } i=r, \ j \neq s, \text{ any } k,t) \\ (\rho_1+\rho_2)\sigma^2, \text{(for } i=r, \ j=s \ k \neq t) \\ \sigma^2, \text{(for } i=r, \ j=s, k=t) \end{cases} \qquad (2.6)$$

where, $k, t = 1, 2, ..., m(i,j), j, s = 1, 2, ..., m(i)$ and $i, r = 1, 2, ..., m$.
Here (2.6), ρ_1 and ρ_2 measure the main cluster and subcluster effects respectively. Note that if it is assumed that $\rho_2 = 0$, then the 3SLR model (2.5) will become equivalent to 2SLR model of the form (2.1). Further note that if it is assumed that each subcluster in a main cluster has only one observation, then at least mathematically one can easily show that (2.6) will become equivalent to that of the two-way error component model.

The general layout of our 3SLR model will be discussed in chapter six, where some diagnostic tests for detecting these cluster and/or subcluster effects are constructed. Monte Carlo methods are used to compare the power performance of these tests with that of the LM tests.

Random Coefficients Model

Note that a model of the form (2.1) is also expressed in terms of a random coefficients model where it is assumed that the regression coefficient, β_k varies from cluster to cluster. A formulation appropriate to the random regression coefficient is given, which is summarised in chapter eight. The formulation of Bhatti (1993) is different from Maddala's (1977, p. 326) variance components model, and Swamy's (1970)[6] and Hsiao's (1986)[7] random coefficients models.

In developing uniform regression models of the form (2.1) and (2.5), the most popular approach of error components models has been used to model survey data. The general popularity of this approach among survey statisticians and panel data econometricians can be attributed to the following factors:

1   They treat huge databases as well as modest ones with equal ease.
2   Estimation and hypothesis testing methods are derived from the classical well known procedures.
3   The problems and difficulties presented remain within the traditional framework, and hence are well understood.
4   The theoretical frontiers are much more explored comparative to other possible approaches.
5   The estimation and hypothesis testing results can be interpreted easily.
6   Most commonly used in econometrics and statistical software packages, which can apply these models with only minor modifications.

The use of error component models in panel data was first suggested by Balestra and Nerlove (1966), and its use in survey data given by Wallace and Hussain (1969), Amemiya (1971), Swamy and Arora (1972), Fuller (1975), Fuller and Battese (1973, 1974), Campbell (1977), Rao and Kleffe (1980),

---

[6] Swamy's (1970) model ignores the equicorrelation coefficient, ρ, within clusters and considers hetroscedasticity. Whereas in the model used in this study, the variance terms of the diagonal elements of the ith clusters are constant within the clusters.
[7] Hsiao (1986) ignores equicorrelation within clusters.

Baltagi (1981), Scott and Holt (1982), Dickens (1990), Baltagi and Raj (1992) and Mátyás and Sevestre (1992), who reviewed the various estimation and testing procedures in the context of panel data. As is noted above, modelling of panel data is very similar to that of the survey data. Therefore, one can apply nearly the same estimation procedures to the 2SLR or (in special situations) 3SLR models that those used in panel data modelling in order to obtain efficient estimates of the unknown parameters.

The hypothesis testing problems associated with the error components models of the type SSMC, SMC, 2SLR, 3SLR and the multistage linear regression models are the main concern of this book. In the next section the theory of PO testing, and the problems and the difficulties involved in developing these tests is summarized.

2.6. OPTIMAL HYPOTHESIS TESTING

In hypothesis testing, one would always prefer to use a UMP test, which maximizes the power curve over the entire parameter space. Unfortunately, the existence of a UMP test does not happen often in practice, perhaps only in special circumstances. When no UMP test exists, it is difficult to decide which is the preferred test, as no single test can dominate in terms of power over the entire parameter space.

Statisticians, econometricians and social scientists are often faced with a high degree of frustration in their respective fields of hypothesis testing because of the limited amount of data available. Furthermore, requisite data cannot always be generated or controlled by a laboratory experiment. Hence with a small amount of information on hand, and when testing for an economic theory (or a hypothesis in any other field), it is important to choose some optimal testing procedure which will provide a powerful test.

Cox and Hinkley (1974, p. 102) consider various techniques for constructing tests of simple null hypothesis, H_0, against a composite alternative hypothesis, H_a, when no UMP test exists. One of the techniques Cox and Hinkley considered using is the most powerful test against a specific alternative $\theta_a \in H_a$ as the test of H_0 against H_a. King (1987) labelled this technique as the PO approach, which is reviewed later in this chapter. In future chapters the method of constructing optimal tests for the SSMC, SMC, 2SLR, 3SLR and the multi-stage models is considered. In the literature, these methods have been shown to work well in a wide variety of testing problems, provided that the specific alternative is chosen carefully. For example, a sequence of papers published by King (1982, 1983, 1985a, 1985b and 1987b) examines the performance of PO tests in testing for first order moving average regression disturbances (MA(1)), first-order autoregressive disturbances (AR(1)) and AR(1) against MA(1) disturbances in the linear regression model with small samples. Further, Evans and King (1985) and King and Skeels (1984) investigate the small sample power properties of PO tests for heteroscedastic disturbances, and joint AR(1) and heteroscedastic regression disturbances respectively.

These studies viewed together allow us to draw the conclusion that certain PO tests have more desirable small sample power properties than others, which includes the LMMP tests. These tests are defined in somewhat more detail later.

Over the past twenty years, the advanced technology provided by the computer revolution has brought enormous benefits not only statisticians, econometricians and other researchers working in the quantitative methods areas. With the ever cheapening cost of computing operations and the greater capacity of computer memory, optimal testing is becoming more viable and popular, and its value is recognized in research work increasingly. No matter whether the hypothesis is simple or composite, one-sided or two-sided, with or without nuisance parameters, advances in computing technology have helped in solving these problems. However, this book is limited to only one-sided testing problems because the cluster effects under tests are mostly positive.

Review of the Theory of Point Optimal Testing

Let's begin by considering a general form of hypothesis testing, i.e. testing
H_0: y has density $f(y, \omega)$, $\omega \in \Omega$

against

H_a: y has density $g(y, \phi)$, $\phi \in \Phi$

in which y is an observed $n \times 1$ vector, ω is a $j \times 1$ vector and ϕ is an $i \times 1$ vector. It is assumed that the possible range of parameter values is determined by all given knowledge, in order to keep the parameter spaces, Ω and Φ, as small as possible.

Generally speaking, there are three different cases in hypothesis testing which are defined as follows:

Case (1): This is a problem of testing simple null against a simple alternative hypothesis, with ω_1 and ϕ_1 fixed and known as parameter values such that $\Omega = \{\omega_1\}$ and $\Phi = \{\phi_1\}$, i.e.,

$$H_0^1 : y \text{ has density } f(y, \omega_1),$$

against

$$H_a^1 : y \text{ has density } g(y, \varphi_1).$$

Therefore, the Neyman-Pearson Lemma (see, Lehmann, 1959, p.65) implies that rejecting H_0^1 for large values of

$$r = g(y, \phi_1) / f(y, \omega_1)$$

is a MP test.

Case (2): This is a problem of testing simple null hypothesis, H_0^1 against the composite alternative, H_a. If r is used as a test statistic, then this test, by construction, is MP in the neighbourhood of $\phi = \phi_1$. The critical value for this test can be found by solving

$$\Pr\{r > r' | y \text{ has density } f(y, \omega_1)\} = \alpha$$

for r' where α is the desired significance level.

Case (3): This is probably the most practical case, and involves testing a composite null hypothesis, H_0, against the composite alternative, H_a. In this situation, it may be difficult to construct an MP test in the neighbourhood of $\phi = \phi_1$. This is because the size of the test and the power of the test may depend on some unknown values of the parameters, i.e., the **Size of Test** is

$$= \Pr\{y \in W | H_0 \text{ is true}\}$$

$$= \int_W pdf_0(y,\omega)dy$$

$$= \alpha(\omega), \text{say.} \tag{2.7}$$

Note that $\alpha(w)$, in (2.7) may depend on the unspecified part of ω, where W is the critical region. Similarly, the **Power of Test** is

$$= \Pr\{y \in W | H_a \text{ is true}\}$$

$$= \int_W pdf_a(y,\phi)dy$$

$$= \beta(\varphi), \text{say.} \tag{2.8}$$

Here also note that $\beta(\phi)$ in (2.8) may depend on ϕ and hence the power of the test may not always be large for some values of ϕ. In equations (2.7) and (2.8), pdf_0 and pdf_a are the probability density functions under the null and alternative hypothesis, respectively.

The standard approach suggested by Lehmann and Stein (1948) is to control the maximum probability of a type I error by the choice of critical value. This is done by solving

$$\sup_{\omega \in \Omega} \Pr\{r > r^* | y \text{ has density } f(y,\omega)\} = \alpha \tag{2.9}$$

for r^* which is the critical value. In general, the choice of r' is greater than r^*. Note that if Ω is a closed set, then

$$\Pr\{r > r^* | y \text{ has density } f(y,\omega)\} = \alpha \tag{2.10}$$

will be true for at least one $\omega \in \Omega$. If there is a value for ω_1 which leads to $r'=r^*$, then the critical regions of the two tests correspond. Thus, it can be said that the test based on r is an MP test in the neighbourhood of $\phi = \phi_1$ of testing H_0 against H_a.

If in the case where the ω_1 value exists, it is also known then constructing a PO test will be much simpler. Therefore, it is important to check for the existence of ω_1 before conducting a PO test, i.e., one has to check whether or not there exists ω_1 value such that (2.10) holds for $\omega = \omega_1$ subject to the constraint that (2.9) also holds. King (1987b, p.175) outlines the method in calculating the left hand side of (2.10) with the help of readily available computer subroutine packages.

The iterative procedure he mentioned can be summarised as the following:

1   Choose a fixed value of ω_1.

2   Solve (2.10) to find r^*, and then check to see if (2.9) holds. If it does hold, then the values of ω_1 and r^* have been found, but if it does not hold, another choice of the ω_1 value is needed by moving it towards the ω values, which caused the violation in (2.9). Then repeat step two.

Unfortunately, in practice this is not always so simple. An ω_1 value sometimes might not exist in a particular testing problem. Lehmann (1959, p.90-94) developed an approach for overcoming the non-existence of an ω_1 value, that is, to reduce a composite null hypothesis to a simple one by considering the weighted average of the distribution over the parameter space. Thus, H_0 is replaced by H_λ with the density function of y is given by ω_1

$$h_\lambda(y) = \int_\Omega f_\omega(y) \, d\lambda(\omega).$$

Lehmann assumes that λ is the degenerate distribution at $\omega = \omega_1$. Under this assumption with the size α, if the most powerful test of a simple null hypothesis, H_λ, against a simple alternative, H_a, has a value of its size less than or equal to α with respect to H_0, then the resultant test is also a PO test of H_0 against H_a. However, if this assumption does not hold, Lehmann's approach is not appropriate. King (1987b, pp. 179-180) and Silvapulle (1991) consider an approximate point optimal (APO) testing approach to solve this problem.

For instance, in case (3) above, a test statistic r with its critical value $r*$ determined by (2.9) is said to be an APO test if $\Pr\{r > r* | H_0^1\}$ is close to α, say within 5% of α. This is because the distribution of H_0 has been changed such that this probability is equal to α, then our test would be a PO test. Therefore, there is a need to look for a ω_1 value, which minimizes the difference between α and $\Pr\{r > R* | y$ has density $f(y, \omega_1)\}$ h,

i.e. $\alpha - \Pr\{r > r* | y$ has density $f(y, \omega_1)\}$. (2.11)

In addition, since (2.11) also depends on ϕ_1, it is worthwhile to consider varying ϕ_1 to minimize the measure of approximation in (2.11) further. King (1989) used the APO approach for testing AR(4) regression disturbances in the presence of AR(1) and testing MA(1) against AR(1) disturbances. Brooks (1993) used an APO test for testing for a single Hildreth-Houck random coefficient against the alternative that the coefficient follows the return to normalcy model.

Invariance Approach and Optimal Tests

Apart from these, one of the most common problems involved in hypothesis testing is the existence of nuisance parameters, which may be present under both the null and alternative hypotheses. The construction of a PO or an APO test requires searching for an appropriate ω_1 value in the parameter space Ω. If the parameter space Ω can be restricted to be as small as possible, then it could be helpful in finding such value for ω_1. Obviously, one should check whether theoretical considerations can indicate some possible restrictions, which can be imposed on the range of various parameters under both H_0 and H_a.

In general, there are two different approaches which help to solve this problem; 'similar and 'invariance approach'. A rejection region of size α is called a 'similar region' of size α if for all values of the nuisance parameters, the probability of rejecting the null hypothesis is exactly equal to α when the null hypothesis is true. Hillier (1987) observed that there is an exact equivalence between similar tests and invariant tests of H0. The 'invariance approach' has considerable success in the application of PO tests because invariance arguments can be used to eliminate nuisance parameters. The reasoning behind this approach is that, if a hypothesis testing problem is invariant to a class of transformations on the observed sample, it is desirable that the test procedure also has this property. For example, as is seen in the subsequent chapters in relation to 2SLR, 3SLR and multi-stage linear regression models that changing of the scale of y and adding a known linear combination of regressors to the rescaled y still preserves the family distributions under both the null and the alternative hypotheses. The conclusion is that the invariance transformation may change the numerical values of the parameters, but the functional form of the likelihood function still remains the same. Ara and King (1993) have shown that the tests based on maximal invariant (i.e. invariant tests) and tests based on the marginal likelihood procedures are equivalent.

As is known in hypothesis testing, for the purpose of test construction a sample space can be divided into two regions, namely, a rejection region and an acceptance region. An invariant test is one for which each pair of points in the sample space that can be related by a transformation, in which both either fall into the rejection region or fall outside the rejection region. Therefore, decisions on the problem of how to partition the sample space can be simplified into choosing which sets of points related by transformations shall be in the rejection region, and which sets shall not.

A convenient summary statistic called a maximal invariant tends to solve the problem of deciding which set of points are to be in a certain region. A maximal invariant exhibits the property of invariance, and every other invariant statistic can be written as a function of it. Hence, the only problem based on the maximal invariant rather than on the observed sample needs to be dealt with. This way the distributions under the null and the alternative hypotheses will have fewer parameters than in the original testing problem. However, the main disadvantage of the invariance arguments is that it is often difficult to find the distributions of the maximal invariant under null and under the alternative hypotheses. Nevertheless, throughout this book, invariance principle will be used in deriving the optimal tests.

Optimality Criterion

A remaining question is, how does one choose a point at which the power is optimized? In the situation when no UMP test exists, Cox and Hinkley (1974, p.102) suggest three different approaches in the choice of points:

1   to pick 'somewhat arbitrary "typical" point' in the alternative parameter space, and use it in the test in order to find a point optimal solution;
2   to remove this arbitrariness choose a 'point' which is very close to the null hypothesis in order to maximize the power locally near the null hypothesis; or
3   to choose a test which maximizes some weighted average of the powers.

Option (1) is known as the 'point optimal' solution, which was discussed earlier in this chapter and reviewed in some detail by King (1987). In the study, Bhatti and King (1990) notes that a class of PO tests is also called 'beta optimal' if its power function is always a monotonic non-decreasing function of the parameter under test, and reaches a predetermined value of the power, say, p_1, most quickly. The concept of beta-optimality was first introduced by Davies (1969), where he optimized power at a value of 0.8, which seems to give good overall power. The construction of the BO test in chapter three optimizes power at a predetermined level of power $p = 0.5$ and 0.8.

In contrast to PO and/or BO tests, option (2) leads us to a locally powerful test, also known as a LB test, a term used throughout this book. The LB test is also optimal in the sense that its power curve has the steepest slope at H_0 of all power curves from tests with the same size.

The LB solution was proposed by Neyman and Pearson (1936), and followed by Efron (1975), King and Hillier (1985) and SenGupta (1987) among others. Let it be assumed that there is interest in testing $H_0: \theta = 0$ based on y, which has been defined earlier as an $n \times 1$ random vector whose distribution has probability density function $f(y|\theta)$ where θ is a $p \times 1$ vector of unknown parameters. When $p=1$, the LB test of $H_0: \theta = 0$ against $H_a^+ : \theta > 0$ is that with critical regions of the form

$$\frac{\partial \ell n f\left(y\mid\theta\right)}{\partial\theta}\Bigg|_{\theta=0} > c_1 \qquad (2.12)$$

where c_1 in (2.12) is a suitably chosen constant (see Ferguson (1967, p.235) and King and Hillier (1985)). King and Hillier (1985) noted that this test is equivalent to the LM1 test based on the square-root of the standard LM test statistic. Against the two-sided alternative $H_a : \theta \neq 0$ when $p =1$, Neyman and Pearson (1936) proposed a test which have the critical regions of the form

$$\frac{\partial^2 f\left(y\mid\theta\right)}{\partial\theta^2}\Bigg|_{\theta=0} > c_2 f\left(y\mid\theta\right)\Big|_{\theta=0} + c_3 \frac{\partial f\left(y\mid\theta\right)}{\partial\theta}\Bigg|_{\theta=0} \qquad (2.13)$$

and which yield LB unbiased (LBU) tests where the constants c_2 and c_3 are chosen so that the critical region has the required size and is locally unbiased. The critical regions of the form (2.13) were labelled as type A regions by Neyman and Pearson in their (1936) paper. Neyman and Pearson also proposed type A_1 tests, which are known as uniformly most powerful unbiased (UMPU) tests. Neyman (1935) showed how to construct type B and type B_1 tests, namely, LBU and UMPU tests, respectively.

For the higher dimension parameter spaces, i.e., where $p \geq 2$, the LBU critical region of $H_0 : \theta = 0$ against $H_a : \theta \neq 0$ is obtained by using Neyman and Pearson (1938) type C LBU regions. These regions have constant power on each of the given family of concentric ellipses in the neighbourhood of the null $\theta = 0$, with this constant power is maximized locally. Isaacson (1951) introduced type D regions to rectify this objection (also see Lehmann, 1959, p.342). Wu (1991) points out that the type D regions are obtained by maximizing the Gaussian curvature of the power function at $\theta = 0$. Furthermore, Wu added, 'in practice, type D critical regions need to first be guessed and then verified'. SenGupta and Vermeire (1986) introduced the class of LMMP unbiased (LMMPU) tests which maximize the mean curvature of the power hypersurface at the null point, i.e. $\theta = 0$ within the class of unbiased tests. Their critical region is of the form

$$\sum_{i=1}^{p}\frac{\partial^2 f\left(y\mid\theta\right)}{\partial\theta_1^2}\Bigg|_{\theta=0} > c_0 f\left(y\mid\theta\right)\Big|_{\theta=0} + \sum_{i=1}^{p} c_i \frac{\partial f\left(y\mid\theta\right)}{\partial\theta_i}\Bigg|_{\theta=0}$$

where the constants c_j, $j = 1,2,...,p$, are chosen such that the test has the nominated size and is locally unbiased.

King and Wu (1990) suggest a multivariate analogue of the one-sided testing problem, i.e. H_0: $\theta = 0$ against $H_a^+ : \theta_1 \geq 0,...,\theta_p \geq 0$ with at least one strict inequality, and show that LMMP test of H_0 against H_a^+ has the critical region

$$s = \sum_{i=1}^{p} \frac{\partial \ell n f \left(y \,|\theta \right)}{\partial \theta_i} > c. \tag{2.14}$$

They also noted that there is always the possibility that one may be able to find such a test of the form (2.14) which is LB in all directions from H_0 in the p-dimensional parameter space. Neyman and Scott (1967) call this property 'robustness of optimality' while King and Evans (1988) refer to such tests uniformly LB (ULB). Like UMP tests, ULB tests may not always exist. In such cases, one may wish to consider a weaker optimality criterion. The aim of the next section is to explore the options of constructing PO, BO, LB and LMMP critical regions in relation to unified regression model for testing cluster/subcluster effects.

2.7. EXISTING TESTS FOR CLUSTER EFFECTS

This section begins by considering articles that construct exact tests for cluster effects i.e., whether the equicorrelation coefficient between observations within a cluster is zero, H_0:$\rho = 0$, against an alternative that it has a positive value, H_a: $\rho > 0$. This section is divided into two parts. The first part reviews articles with the stacked random vectors of different observations which follow SSMC and SMC distributions. The second part of this section whereas will concentrate on testing for cluster and subcluster effects in the context of the multi-stage linear regression model, when disturbances follow only SMC distributions.

SSMC and SMC Models

Sampson (1976, 1978) has considered applications of the SSMC distribution and the estimation of its equicorrelation coefficient, ρ. William and Yip (1989) whereas used the idea of ancillary statistics on the two-stage SSMC distribution model, and found that the tests and the confidence intervals based on ancillary statistics were conditionally optimal. In this section, the main concern is to survey articles, which have considered exact optimal tests so that the power of the new and the existing exact tests can be compared.

SenGupta (1988) proposed a LB test for non-zero values of the ρ in a two-stage SSMC distribution model (2.3). SenGupta showed that in the context of (2.3), the likelihood ratio test that ρ takes a given value has a number of theoretical and practical shortcomings. SenGupta constructed the LB test of H_0: $\rho = 0$ against H_a: $\rho > 0$ and found that his tests do not share these shortcomings. This test instead rejects H_0 for small values of

$$d = \sum_{i=1}^{m} \sum_{j=1}^{k} y_{ij}^2 - \sum_{i=1}^{m} \left(\sum_{j=1}^{k} y_{ij} \right)^2. \tag{2.15}$$

Empirical power comparisons of these tests showed the superiority of the LB test. A LB test whose statistic is of the form d in (2.15) can be viewed as a test which optimizes power in the neighbourhood of the null hypothesis.

In chapters three and four PO tests are considered which optimize power at a predetermined point, away from the null hypothesis. The main findings of these chapters are that for the testing problem in the context of SSMC and SMC distributions, the point optimal tests are BO, and UMP invariant (UMPI), respectively. An extension of the two-stage SSMC and SMC models to multistage model is considered and a class of new optimal tests has been proposed.

Multi-stage linear regression models

Honda (1991) proposes a standardised test for the two-way error component model and found that the size of his test improves the accuracy for the small sample sizes. Baltagi, Chang and Li (1991) consider one-way and two-way error component models and propose various tests on the variances-covariances of the individual and time effects. Mátyás and Sevestre (1992) review some of the tests suggested by Honda and Baltagi et al, which were associated with the panel data models. Others who have consider error component models are Honda (1985) and Moulton and Randolph (1989). Honda, Moulton and Randolph analyses the behaviour of the LM1 tests, and demonstrates that it is more powerful and more robust to non-normal error terms than the LM2 test. King and Hillier (1985) showed that the LM1 test is LBI, with these properties suggesting the usefulness of the LM1 test in the context of modelling survey data. The main objective of this section is to survey some of those articles which consider testing cluster/subcluster effects associated with the 2SLR and the 3SLR models.

Deaton and Irish (1983) consider the 2SLR model of the form (2.1). They discuss the problem of testing for cluster effects and suggest the use of the LM1 (score) test. This test rejects $H_0{:}\rho = 0$ for large values of

$$g = \hat{\rho} \left(\frac{1}{2} \sum_{i=1}^{m} m(i) \left[m(i) - 1 \right] \right)^{\frac{1}{2}}$$

(2.16)

where,

$$\hat{\rho} = \sum_{i=1}^{m} \sum_{j=1}^{m(i)} \sum_{k<j} \hat{u}_{ij} \hat{u}_{ik} \left/ \left[\frac{1}{2} \sum_{i=1}^{m} m(i) \left\{ \left(m(i) - 1 \right) \right\} \hat{\sigma}^2 \right] \right.,$$

in which $\hat{u}_{ij}$ are OLS residuals and $\hat{\sigma}^2$ is the standard OLS estimator of the disturbance variance. It is worth noting that asymptotically g has a standard normal distribution under H_0. The two-sided Lagrange multiplier test rejects H_0 for large values of g^2, which has an asymptotic $\chi 2$ distribution with one degree of freedom under the null hypothesis.

The g test in (2.16) requires the knowledge of the cluster structure for its application. Deaton and Irish (1983) have pointed out that this may not always be available, with cluster information often removed from survey data before release, although the original cluster ordering is retained. Their suggest that in such cases one may have to resort to the use of the Durbin-Watson test, which is attractive, especially since this test is available in most standard output regression packages.

King and Evans (1986) also consider the 2SLR model with the same testing problem, and discuss the small sample optimal power properties of the LM1, LM2 tests and the DW test. King and Evans empirically compare the powers of these tests and found that the LM1 test is LBI. In chapter five when dealing with the 2SLR model this testing problem is considered, and the construction of the POI, LMMPI and King's (1981) modified DW test is worked out. It is found that the power of the POI test is marginally superior to that of the existing tests in the literature.

Others who consider the 2SLR model are Wu et al. (1988) and Rao et al. (1993), who allow negative and/or unequal equicorrelation in their models. Wu et al. (1988) develop a standard F-test for testing H_0: $C\beta = b$ whereas Rao et al. (1993) develop a generalised least square (GLS) F test. The power and size comparison between Wu et al's F test and Rao's et al's GLS F test demonstrates the superiority of the GLS F test. Wu and Bhatti (1994) consider the 3SLR model and suggest the use of POI and LMMPI tests for testing cluster effects and misspecification in regression models based on survey data. This book follows Wu and Bhatti's procedure of detecting cluster/subcluster effects from the SSMC, SMC, 2SLR, 3SLR and finally the multistage Linear regression models in the subsequent chapters.

2.8. CONCLUDING REMARKS

This chapter reviewed other scholars' work to understand the issues and problems connected with the use of sample surveys in regression analysis. Most of the scholars have considered the 2SLR model, which at times inappropriate in real life situations. Therefore, their work was extended to a multi-stage model, by incorporating cluster, subcluster and multicluster effects in the error terms. The choice of using model-based approach in the analysis of survey data was explained, and a unified regression model for survey and panel data was developed. While these are the preliminary concepts a proper understanding of them is necessary before embarking on the task of developing the optimal tests. The theory of PO testing was also summarized in this section.

Articles that construct exact tests for testing cluster effects were reviewed so that the tests used in this book could stand for comparison later on. The literature survey of the PO tests revealed that they performed pretty well in other situations in terms of their size and power properties, compared to asymptotic tests. This suggests that it is worthwhile to explore the power properties of the optimal tests, like PO, BO, LB and LMMP, in testing these cluster/subcluster effects. Therefore, in the subsequent chapters' tests on SSMM, SMC, 2SLR, 3SLR and multi-stage linear regression models are applied.

3. TESTS FOR CLUSTER EFFECTS OF THE SSMC MODEL

3.1. INTRODUCTION[1]

The literature survey in chapter two revealed that a model, which follows the SSMC distribution, has wide applications in many fields of human endeavour, including econometrics, statistics, biological and physical sciences. The main aim of this chapter is to consider several aspects of hypothesis testing problems associated with models based on two- and three-stage SSMC distributions. Discussion will be concentrated on developing optimal tests, which are touched upon in the previous chapter, with this chapter will also explore the possibility of constructing a PO test for testing the subcluster effects in the presence of main cluster equicorrelation.

This chapter is structured as follows. In the following section of this chapter, the two-stage SSMC model is introduced and the BO test is constructed. The following section also discusses the LB test, two versions of BO tests and the PE, and it provides tables of selected critical values. An empirical power comparison involving LB and BO tests for the case of two-stage SSMC distributions is also reported, with the application of the LMMP test to more complicated models is given, where means and variances are known but cluster effects are unknown and vary from cluster to cluster. The main interest of that section is to test whether these varying cluster effects are equal to zero against the alternative where they have some positive values,

---

[1] A paper presenting some of the findings reported in this chapter were published in the Australian Journal of Statistics, Vol. 32(1), p. 87-97, Bhatti and King (1990).

i.e., $H_0{:}\rho_i = 0$, $i = 1,...,m$ against $H_a{:}\rho_i \geq 0$, $i = 1,...,m$, with at least one inequality being a strict inequality. How LMMP and PO critical regions can be constructed in the context of the three-stage SSMC distribution model is also examined and there is discussion on more complicated matters such as testing for subcluster effects in the presence of main cluster effects in the case of the three-stage cluster model. The final section of this chapter will contain some concluding remarks.

3.2. TWO-STAGE SSMC MODEL AND THE TESTS

Let y be an observed $k \times 1$ random vector with a SSMC distribution, i.e.

$$y \sim N\left(0, \sum(\rho)\right),$$

where the matrix $\sum(\rho)$ is already been defined by (2.3) in chapter two. Furthermore, note that the simplest form of (2.3) can be written as

$$\sum(\rho) = \left[(1-\rho)I_k + \rho E_k\right],$$

(3.1)

where I_k is the $k \times k$ identity matrix, and E_k is the $k \times k$ matrix with all elements equal to unity. Then the inverse of (3.1) is given by

$$\sum(\rho)^{-1} = (1-\rho)^{-1}\left[I_k - \rho\left\{1 + (k-1)\rho\right\}^{-1} E_k\right].$$

(3.2)

Hence, the density function of y when $\sum(\rho)$ is non-singular can be expressed as

$$f(y,\rho) = \frac{1}{(2\pi)^{k/2}\left|\sum(\rho)\right|^{1/2}} \exp\left[-\frac{1}{2}\left\{\frac{\left(\sum y_i^2\right)}{(1-\rho)} - \frac{\rho\left(\sum y_i\right)^2}{\left\{1+(k-1)\rho\right\}(1-\rho)}\right\}\right],$$

$$-\infty > y_i > \infty, i = 1, ..., k, \text{ provided } \frac{-1}{(k-1)} < \rho < 1. \qquad (3.3)$$

The restriction on the range of ρ is imposed to avoid the cases where $\sum(\rho)$ is singular. For larger values of k, this range effectively becomes $0 \le \rho < 1$. The remainder of this chapter is restricted to nonnegative values of ρ.

Let $y_1, y_2, ..., y_m$ denote random samples of observed values from each of the m clusters from the distribution given in (3.3), and let Y denote the $n \times 1$ stacked vector of these independent random vectors where $n = mk$. This can be expressed in matrix form as

$$Y = \begin{pmatrix} y_1 \\ y_2 \\ \vdots \\ y_m \end{pmatrix} = \begin{pmatrix} y_{11} \\ y_{12} \\ \vdots \\ y_{1k} \\ \vdots \\ \vdots \\ y_{m1} \\ y_{m2} \\ \vdots \\ y_{mk} \end{pmatrix}.$$

where the distribution of Y would be of the form

$$Y \sim N\left(0, \Delta(\rho)\right)$$

where,

$$\Delta(\rho) = I_m \otimes \sum(\rho)$$
$$= (1 - \rho)I_n + \rho D, \qquad (3.4)$$

$\otimes$ denotes the Kronecker product and D is the $n \times n$ cluster diagonal matrix of E_k's

$$D = I_m \otimes E_k$$

$$= \begin{pmatrix} E_k & 0 & \cdots & 0 \\ 0 & E_{k\cdots} & & 0 \\ . & . & . & . \\ . & . & . & . \\ . & . & & . \\ 0 & \cdots & & E_k \end{pmatrix}. \tag{3.5}$$

Furthermore, the inverse of $\Delta(\rho)$ is expressed as

$$\Delta^{-1}(\rho) = (1-\rho)^{-1} \left[I_n - \rho \{1 + (k-1)\rho\}^{-1} D \right]. \tag{3.6}$$

Note that the error covariance matrix in (3.4) involves only one unknown parameter, ρ, which ranges from 0 to +1 under (3.3). In the next subsection of this chapter, the BO test for detecting the presence of equicorrelation within clusters will be constructed.

Beta-Optimal Test

The problem of interest is one of testing

$$H_0 : \rho = 0$$

against the one-sided alternative

$$H_a : \rho > 0.$$

Under H_0, the density function of Y is given by

$$f_0(Y) = (2\pi)^{-n/2} \exp\left\{\frac{-1}{2} Y'Y\right\}$$

while under H_a it is given by

$$f_a(Y; \rho) = (2\pi)^{-n/2} |\Delta(\rho)|^{-1/2} \exp\left[\frac{-1}{2} Y'\Delta^{-1}(\rho)Y\right].$$

It is easier to first consider the simpler problem of testing H_0 against the simple alternative hypothesis $H_1 : \rho = \rho_1 > 0$ where ρ_1 is fixed and a known value of the parameter under H_0. The Neyman-Pearson Lemma implies that the critical region as

$$|\Delta(\rho_1)|^{-1/2} \exp\left[\frac{-1}{2} Y'\Delta^{-1}(\rho_1)Y + \frac{1}{2} Y'Y\right] \geq c$$

or equivalently

$$r(\rho_1) = Y'\left(\Delta^{-1}(\rho_1) - I_n\right)Y < c_\alpha \tag{3.7}$$

which is the most powerful test, where c_α is an appropriate critical value.

It should be noted that under (3.6), one can express $r(\rho_1)$ in (3.7) as a simple series of sums as follows, i.e.

$$r(\rho_1) = (a*-1)Y'Y - a*b*Y'DY$$

$$= (a*-1)\sum_{i=1}^{m}\sum_{j=1}^{k} y_{ij}^2 - a*b*\left(\sum_{i=1}^{m}\sum_{j=1}^{k} y_{ij}\right)^2,$$

where

$$a* = \frac{1}{1-\rho_1}, \quad b* = \frac{\rho_1}{1+(k-1)\rho_1}.$$

In order to compute c_α, let $A = (\Delta^{-1}(\rho_1)\text{-}I_n)$, and let λ_i, $i=1,...,n$ be the eigenvalues of A, and let P be an $n \times n$ orthogonal matrix such that

$$PAP' = \begin{pmatrix} \lambda_1 & 0 & . & . & . & . & 0 \\ 0 & \lambda_2 & . & . & . & . & 0 \\ . & & . & . & . & & . \\ . & & . & . & . & & . \\ . & & . & . & & . & \\ 0 & . & . & . & . & . & \lambda_n \end{pmatrix}.$$

Then under H_0

$$\Pr\left[r(\rho_1) < c_\alpha \right]$$

$$= \Pr\left[Y'AY < c_\alpha \right],$$

$$= \Pr\left[\sum_{i=1}^{n} \lambda_i \xi_i^{\,2} < c_\alpha \right]$$

$$= \Pr\left[\frac{\rho_1}{1-\rho_1} \chi^2_{m(k-1)} - \frac{\rho_1(k-1)}{\{1+(k-1)\rho_1\}} \chi^2_m c_\alpha \right] \qquad (3.8)$$

where $\xi = (\xi_1, \xi_2, ..., \xi_n) \sim N(0, I_n)$ and χ^2_j denotes a chi-squared random variable with j degrees of freedom. The last equality of (3.8) can be obtained by noting that the eigenvalues of D are k and zero, with multiplicities of m and $m(k-1)$ respectively. This together with (ii) and (iii) of lemma 1, in appendix 3A implies that the eigenvalues of

$$\left(\Delta^{-1}(\rho_1) - I_n \right)$$

are

$$\frac{-\rho_1(k-1)}{\left[1+(k-1)\rho_1\right]} \text{ and } \frac{-\rho_1}{(1-\rho_1)}$$

with multiplicities m and $m(k-1)$, respectively.

In order to find c_α such that (3.8) equals the required significance level, α, (3.8) can be evaluated using either Koerts and Abrahamse's (1969) FQUAD subroutine or Davies' (1980) algorithm. Alternatively, in (3.8), $r(\rho_1)$ is expressed as the weighted difference of two independent chi-squared random variables, so that its probability density function is given by SenGupta (1987, Theorem 3).

For the wider problem of testing H_0 against H_a, the test based on $r(\rho_1)$ is most powerful at $\rho = \rho_1$ and is therefore a PO test. A central question is how should ρ_1 be chosen? Strategies for choosing the point at which a PO test optimizes power are discussed in King (1987). One approach, which King suggests, is to choose a value of ρ_1 arbitrarily. Another approach is to take the limit of $r(\rho_1)$ tests as ρ_1 tends to zero, which results in SenGupta's LB test. This chapter will follow King's favoured approach of choosing ρ_1 arbitrarily, so that the resultant test's power is optimised at a predetermined level of power denoted ρ_1. There seems little point in optimising power when it is very low as the LB test does, or when it is one or nearly one. Optimizing power at middle power value say 0.5 or 0.8 is favoured more so. In order to do this, the power of the test needs to be calculated readily.

Consider the Cholesky decomposition of $\Delta(\rho)$, namely

$$\Delta(\rho) = TT'$$

where T is a $n \times n$ nonsingular, lower triangular matrix. Then

$$\Delta^{-1}(\rho) = \left(T^{-1}\right)' T^{-1}$$

and under H_a

$$z = T^{-1}Y \sim N\left(0, I_n\right).$$

Because of the cluster diagonal nature of the $\Delta(\rho)$, T is also cluster diagonal of the form

$$T = \begin{pmatrix} S_k & 0 & \cdots & & 0 \\ 0 & S_k & \cdots & & 0 \\ \cdot & & \cdot & & \cdot \\ \cdot & & & \cdot & \cdot \\ \cdot & & & & \cdot \\ 0 & & \cdot & \cdot & S_k \end{pmatrix}$$

where S_k is a $k \times k$ lower triangular matrix whose elements are obtained by

$$s_{ij}(\rho) = \rho \sqrt{\frac{(1-\rho)}{\left[1+(i-2)\rho\right]\left[1+(i-1)\rho\right]}}$$

and

$$s_{ii}(\rho) = \sqrt{\frac{(1-\rho)\left[1+(i-1)\rho\right]}{\left[1+(i-2)\rho\right]}} \tag{3.9}$$

for

$i = 1,2,...,k$; $j = 2,3,...,k$ and $j > i$.

Thus for any critical value c_α the power of the critical region $r(\rho_1) < c_\alpha$ is

$$\Pr\left[r(\rho_1) < c_\alpha \middle| Y \sim N\left(0, \Delta(\rho)\right)\right]$$

$$= \Pr\left[z'T'\left(\Delta^{-1}(\rho_1) - I_n\right)Tz < c_\alpha \right]$$

$$= \Pr\left[\sum_{i=1}^{n} \lambda_i \xi_i^2 < c_\alpha \right], \tag{3.10}$$

where λ_i, $i = 1,2,...,n$, are the eigenvalues of

$$T'\left(\sum^{-1}(\rho_1) - I_n\right)T \text{ and } \xi_i \sim N(0,1).$$

This follows from lemma 2 of appendix 3A, that the eigenvalues of $T'\left(\Delta^{-1}(\rho_1) - I_n\right)T$ are also the eigenvalues of

$$\left(\Delta^{-1}(\rho_1) - I_n\right)\Delta(\rho)$$

$$= \frac{\rho_1(1-\rho)}{(1-\rho_1)}I_n + \frac{\rho_1\left[\rho\{(\rho_1-1)(k-1)+1\}-1\right]}{\left[(1-\rho_1)\{1+(k-1)\rho_1\}\right]}D$$

$$= aI_n + bD$$

say. The eigenvalues of the latter matrix are

$$a = \frac{\rho_1(1-\rho)}{(1-\rho_1)} \tag{3.11}$$

with multiplicity of $mk-m$ and

$$a + bk = \frac{-\rho_1(k-1)\{1+(k-1)\rho\}}{\{1+(k-1)\rho_1\}} \tag{3.12}$$

with multiplicity of m. Thus $r(\rho_1)$ in (3.10) can also be expressed as a weighted difference of two independent chi-squared random variables, so (3.10) can be evaluated using any of the methods outlined for calculating (3.8). In the special case in which power is being evaluated at $\rho = \rho_1$, (3.11) and (3.12) reducing to ρ_1 and $-(k-1)\rho_1$, respectively so that (3.10) can be written as

$$\Pr\left[\rho_1 \chi^2_{m(k-1)} - (k-1)\rho_1 \chi^2_m < c_\alpha\right]$$

$$= \Pr\left[\chi^2_{m(k-1)} - (k-1)\chi^2_m < c_\alpha / \rho_1\right]. \tag{3.13}$$

Observe that both (3.11) and (3.12) decline in value as ρ increases. Given (3.10), this means that the test's power increases as ρ increases, which implies that the test is a BO test. As noted by King (1987b, p.197-198), a PO test is BO if its power function is always a monotonic non-decreasing function of the parameter under test. This concept of beta-optimality was first introduced by Davies (1969), who suggested that the value of ρ_1 should be 0.8.

Given the desired level of significance, α, and the level of power, ρ_1, at which power is optimized, then ρ_1 and the associated critical value c_α can be found in two steps using (3.13) as follows:

1 Solve $\Pr\left[\chi^2_{m(k-1)} - (k-1)\chi^2_m < c_\alpha / \rho_1\right] = p_1$ for c_α / ρ_1

2 Given the solution of the ratio of c_α / ρ_1 from step (1) above, determine c_α and ρ_1 by solving

$$\Pr\left[r(\rho_1) < c_\alpha\right] = \alpha$$

.

The remainder of this section will denote $r(\rho_1)$ test as the r_{p_1}.

Selected one and five percent significance points, c_α, and their associated ρ_1 values for the $r_{0.5}$ (i.e., $\rho_1 = 0.5$) and $r_{0.8}$ (i.e., $\rho_1 = 0.8$) tests are tabulated in tables 3.1 and 3.2 of appendix 3C, respectively. These tests were calculated using a FORTRAN version of Davies' (1980) algorithm, with these calculations made for vectors of size $k = 2,3,...,10$ and for sample of sizes $m = 5,10,15$ and 25. In recent days, this computation work can be done easily via variable subroutines.

3.3. POWER COMPARAISON OF BO, LB TESTS & PE

PO tests can be used to trace out the maximum attainable PE for a given testing problem, which in this case can be done by evaluating the power of the $r(\rho_1)$ test at $\rho = \rho_1$ over a range of ρ_1 values. The PE provides a clear benchmark against which test procedures can be evaluated. If the power of a test is always found to be close to the PE, then it can be argued that the test is an approximately UMP test. An example of such a finding is given by Shively (1988).

It is of interest to compare the power curves of SenGupta's LB test with those of the two versions of the BO tests, $r_{0.5}$ and $r_{0.8}$, to see whether one of these tests is close enough to the PE to be called an approximately UMP test.

SenGupta uses Efron's (1975) criterion of statistical curvature, γ_θ, to measure the small sample performance of the LB test. Based on 'very rough calculations', Efron (1975, p.1201) suggested that

$$\gamma_{\theta_0}^2 \geq 1/8 \qquad\qquad (3.14)$$

is a 'large' value where θ_0 is the value of the parameter under the null hypothesis. In such a case, it is reasonable to question the use of a LB test. For the testing problem defined in the previous section entitled 'Beta-optimal test', SenGupta found that (3.14) is equivalent to $mk \leq 64$. How good is Efron's rule in such a case, dealing with SSMC distributions?

With such thoughts in mind, the PE is computed and compared with the powers of the LB and the two versions of the BO tests, i.e., $r_{0.5}$ and $r_{0.8}$, at the five percent level of significance. Calculations of these powers at $\rho = 0.05$, 0.1, 0.2, 0.3, .., 0.9 for $m = 10, 15, 25$ and $k = 2, 3, 4, 6, 10$ are done. These calculations are obtained by using Davies' (1980) algorithm, selected results are tabulated in tables 3.3, 3.4 and 3.5 of appendix 3C. The values for $\rho = 0.7$ and 0.9 have been omitted because, especially for large values of m and k, they are very similar to those for $\rho = 0.8$.

These results demonstrate that the PE and the powers of all tests increase as k, m and ρ increase, while other variables remain the same. As expected, from the three tests, the LB test is most powerful for ρ values associated with small PE probabilities, whereas the $r_{0.5}$ test is most powerful for ρ values associated with middle PE probabilities, and the $r_{0.8}$ test is most powerful for ρ

values associated with large PE probabilities. Particularly for larger k and m values, the PE and all three power curves generally reach a value of one as ρ increases. The PE reaches this maximum value first, followed by the power curves of the $r_{0.8}$, $r_{0.5}$ and LB tests respectively.

More importantly, for each combination of k and m values, the LB test always has the largest maximum power deviation below the PE of the three tests. For example, for $k = 2$ and $m = 10,15,25$, this maximum power difference is 0.246, 0.162 and 0.095, respectively. In contrast, the largest maximum power difference between the $r_{0.5}$ test and the PE is never greater than 0.031, while that for the $r_{0.8}$ test is never greater than 0.027. On the basis of these results it can be said that the $r_{0.5}$ and $r_{0.8}$ tests are approximately UMP, at least at the 5% level. If either k or m increases, the maximum power difference from the PE decreases as either k (or m) increases, while other variables remain the same.

Efron's rule of questioning the use of a LB test when (3.14), or equivalently $mk \leq 64$ holds, appears to work well in this situation. Maximum deviations from the PE when $mk \leq 64$ range from 0.246 to 0.071, while for $mk > 64$ they range from 0.057 to 0.015. There seems to be a tendency for the rule to work better for smaller m values.

There is the question of which of the r_{p_1} tests is better. While the $r_{0.5}$ test has larger maximum deviations from the PE, the $r_{0.8}$ test has larger maximum percentage deviations; this is because the $r_{0.5}$ test has increased power for lower levels of power, while the $r_{0.8}$ test is relatively more powerful for higher levels. The differences between the two tests are not great, with the choice of test, therefore boiling down to a choice between extra powers at lower or higher values of ρ.

3.4. A MORE COMPLICATED MODEL AND THE LMMP TEST

The error covariance matrix in (3.4) involves only one parameter, ρ, but in practice it may be that the structure of the correlated observations changes from cluster to cluster. Hence the cluster effects also vary from cluster to cluster or from industry to industry. Broadly speaking, one might view the error as being the sum of two components, with the first being a disturbance for each cluster that is constant for each observation within the cluster. The second is a disturbance for individual observations. If the variance of this latter disturbance varies from cluster to cluster then one can obtain a correlation coefficient that also varies from cluster to cluster. For example, in studying the consumption patterns of different income groups in a particular locality, one can find observations within each group that may follow a similar pattern, but in their consumption behaviour may vary from one to another. Therefore, it is reasonable to assume that the degree of intragroup correlation differs from one cluster to another cluster; hence, this would result in different variances and covariances, which leads to different equicorrelation coefficients for each income group.

In literature, these situations have been explored by many authors, with the applications of various tests for the equality of means, variances-covariances and equicorrelation coefficients found in Wilks (1946), Votaw (1948), Geisser (1963), Selliah (1964) Srivastava (1965) and Bhatti (2004). The objective of the next section is to construct one-sided test for testing whether these varying intra cluster correlation are equal to zero for different groups, and even for different subgroups, against the alternative that they have some positive values, which will be done by using Bhatti's (1992) LMMP and PO tests.

Locally Most Mean Powerful Test

The discussion in this section begins by noting that if different groups have different intra cluster correlation coefficients, then the covariance matrix $\Delta(\rho)$, in (3.4) can be expressed as

$$\Delta(\rho) = \begin{bmatrix} \Sigma(\rho_1) & 0 & \cdots & 0 \\ 0 & \Sigma(\rho_2) & \cdots & 0 \\ \vdots & \vdots & \ddots & \vdots \\ 0 & \cdots & \cdots & \Sigma(\rho_m) \end{bmatrix} \tag{3.15}$$

$$= \bigoplus_{i=1}^{m} \Sigma(\rho_i),$$

which is block-diagonal with submatrices of the form

$$\sum(\rho_i) = (1 - \rho_i)I_k + \rho_i E_k \tag{3.16}$$

for $i=1,2,...,m$, clusters, where $\rho = (\rho_1,\rho_2,...,\rho_m)'$ and I_k and E_k already defined in the previous sections of this book.

The problem of interest is one of testing

$$H_0 : \rho_1 = \rho_2 = ... = \rho_m = 0$$

against

$$H_a : \rho_1 \geq 0, \rho_2 \geq 0, ..., \rho_m \geq 0,$$

with at least one inequality being a strict inequality, which is a m-dimensional hypothesis testing problem. Therefore King and Wu's (1990) theorem to construct a LMMP test for testing H_0 against H_a can be used in this case that the distribution of Y is

$$Y \sim N(0, \Delta(\rho)),$$

where the $n \times n$ matrix $\Delta(\rho)$ is given by (3.15) with submatrices (3.16), and is such that $\Delta(0) = I_n$. The density function of Y with ρ being a $m \times 1$ vector is given by

$$f(Y | \rho) = (2\pi)^{-n/2} |\Delta(\rho)|^{-1/2} \exp\left[-\frac{1}{2}Y'\Delta^{-1}(\rho)Y\right] \tag{3.17}$$

where

$$\Delta^{-1}(\rho) = \begin{bmatrix} \Sigma^{-1}(\rho_1) & 0 & \cdots & 0 \\ & & & \cdot \\ 0 & \Sigma^{-1}(\rho_2) & & \cdot \\ \cdot & & & \cdot \\ \cdot & & & \cdot \\ \cdot & & & \cdot \\ \cdot & & & \cdot \\ 0. & \cdots & 0 & \Sigma^{-1}(\rho_m) \end{bmatrix}, \qquad (3.18)$$

such that the inverse of each of $\Sigma(\rho_i)$ is given by (3.2), i.e.

$$\Sigma^{-1}(\rho_1) = (1-\rho_i)^{-1}\left[I_k - \rho_i\left\{1+(k-1)\rho_i\right\}^{-1} E_k \right].$$

From (2.13) in chapter two the LMMP test of H_0 against H_a, based on the density (3.17), is given by the critical region

$$\sum_{i=1}^{m} \left.\frac{\partial \ell nf(Y \mid \rho)}{\partial \rho_i}\right|_{\rho=0} > c, \qquad (3.19)$$

where c is the appropriate constant chosen to give the required significance level of our test.

Thus, from (3.17), (3.18) and (3.19), the LMMP test is based on critical regions of the form

$$\sum_{i=1}^{m} -\frac{1}{2}\left.\frac{\partial \ell n|\Delta(\rho)|}{\partial \rho_i}\right|_{\rho=0} - \sum_{i=1}^{m}\frac{1}{2}\left.\frac{\partial\left(Y'\Delta^{-1}(\rho)Y\right)}{\partial \rho_i}\right|_{\rho=0} > c.$$

The first summation on the left is simply a constant, whilst evaluating the second summation gives the critical region

$$-\frac{1}{2}c_1 - \frac{1}{2}\sum_{i=1}^{m} Y'A_iY > c,$$

or equivalently,

$$d = \sum_{i=1}^{m} d_i = \sum_{i=1}^{m} Y'A_i Y = Y'AY \le c_2, \tag{3.20}$$

where $A = \sum_{i=1}^{m} A_i$ and c_2 is a suitably chosen constant. Note that the matrices A_i can be obtained from (3.18) as follows, i.e.

$$A_i = \frac{\partial \Delta^{-1}(\rho)}{\partial \rho_i}\Big|_{\rho=0}, i = 1, 2, ..., m$$
$$= \{a_{i, j\ell}\}$$

where

$$a_{i, j\ell} = \begin{cases} -1, & \text{for } (i-1)k < j < ik+1, \\ & (i-1)k < \ell < ik+1 \\ & \text{and } j \ne \ell; \\ 0, & \text{otherwise} \end{cases} \tag{3.21}$$

for $i=1,2,...,m$ and $j, \ell =1,2,...,n$. That is, the A'_is are cluster diagonal matrices, whose cluster are m x m and all but the ith diagonal cluster, which is $(I_k - E_k)$ that are zero. Thus by (3.16) and (3.18),

$$A = I_m \otimes (I_k - E_k)$$
$$= I_n - D, \tag{3.22}$$

where D is defined in (3.5). It is worth noting that (3.20) is a LB test of H_0 in the direction $\rho_1 = \rho_2 = ... = \rho_m > 0$, which is not unusual. A similar example of this type of result can be found while dealing with quarter-dependent fourth-order autocorrelated models (see King and Wu, 1990, p.17-18).

Moreover, under (3.22), d in (3.20) can also be expressed in a simplified form as follows,

$$d = Y'AY$$

$$= Y'(I_n - D)Y$$

$$= \sum_{i=1}^{m}\sum_{j=1}^{k} y_{ij}^2 - \sum_{i=1}^{m}\left(\sum_{j=1}^{k} y_{ij}\right)^2, \tag{3.23}$$

which is in the form of 2.14 of chapter two (i.e., Sen Gupta's LB test). Hence Sen Gupta's LB test is also a LMMP test.

Furthermore, one can note that in (3.23), d is a quadratic form in normal variable and therefore its critical value c_2 may be obtained by evaluating probabilities of the form

$$\Pr\left[d \le c_2\right]$$

$$= \Pr\left[Y'(I_n - D)Y \le c_2\right]$$

$$= \Pr\left[\sum_{i=1}^{n} \lambda_i \xi_i^2 \le c_2\right]$$

$$= \Pr\left[\chi_{m(k-1)}^2 - (k-1)\chi_m^2 \le c_2\right] \tag{3.24}$$

where $\xi = (\xi_1,...,\xi_n)' \sim IN(0,I_n)$ and λ_i, $i=1,...,n$ are the eigenvalues of the matrix (I_n-D), namely 1 and $(1-k)$ with multiplicities of $m(k-1)$ and m, respectively.

3.5. THREE-STAGE SSMC MODEL AND THE TESTS

Theory

This section extends the two-stage SSMC distributions model considered earlier to a more general situation where observations are obtained from a three-stage (or higher-stage) cluster design. Suppose y_{ijk} denotes a kth (third-stage) selected variate value from a jth (second-stage) subcluster of the ith (first-stage) main cluster. Let Y denote the $n \times 1$ random vector, which is formed by stacking different observations for each subcluster, then stacking different subclusters for each cluster, and finally stacking for different clusters.

There are a total of n observations which are sampled from m first stage clusters (or groups), with m(i) subcluster from the ith cluster and with m(i,j) observations from the jth subcluster of the ith cluster, such that

$$n = \sum_{i=1}^{m} \sum_{j=1}^{m(i)} m(i, j).$$

The random vector Y is said to follow the three-stage SSMC distribution model[2] if

$$Y \sim N\left(0, \Omega\left(\rho_1, \rho_2\right)\right) \tag{3.25}$$

where $\Omega\left(\rho_1, \rho_2\right) = \bigoplus_{i=1}^{m} \Omega_i\left(\rho_1, \rho_2\right)$ is cluster diagonal, with submatrices

$$\Omega_i\left(\rho_1, \rho_2\right) = \left(1 - \rho_1 - \rho_2\right) I_{m_i} + \rho_1 E_{m_i} + \rho_2 \bigoplus_{j=1}^{m(i)} E_{m(i,j)} \tag{3.26}$$

where $m_i = \sum_{j=1}^{m(i)} m\,(i, j)$, I_{m_i} is an $m_i \times m_i$ identity matrix, E_{m_i} and $E_{m(i,j)}$ are $m_i \times m_i$ and $m(i,j) \times m(i,j)$ matrices, respectively one.

---

[2] This is a special case of the 3SLR model discussed at (2.5).

Although unequal cluster and subcluster sizes can easily be used in this model, for the sake of convenience, in exposition, the rest of this chapter will consider only balanced cases, where data are balanced over subclusters. Therefore, it is assumed that $T=m(i,j)$, $N_i = m(i)$, for $i=1,...,m$ and $j=1,...,N_i$, then (3.26) can be written as

$$\Omega_i\left(\rho_1,\rho_2\right) = \left(1-\rho_1-\rho_2\right)\left(I_{N_i} \otimes I_T\right) + \rho_1\left(e_{N_i}e'_{N_i} \otimes e_T e'_T\right) + \rho_2\left(I_{N_i} \otimes e_T e'_T\right) \tag{3.27}$$

where e_T and e_{N_i} are $T \times 1$ and $N_i \times 1$ vectors of ones.

Locally Most Mean Powerful Test

The error covariance matrix in (3.26) or equivalently in (3.27) involves two unknown parameters, i.e., the intra cluster correlation coefficient, ρ_1 and the intra subcluster correlation coefficient, ρ_2.[3] As mentioned earlier in chapter one and two, ignoring such an intracluster or intrasubcluster correlation coefficient may result in seriously misleading estimation and hypotheses testing results based on inefficient inference procedures.

Therefore, the problem of interest in this section is to test

$$H_0 : \rho_1 = \rho_2 = 0 \tag{3.28}$$

against

$$H_a : \rho_1 \geq 0, \rho_2 \geq 0, \text{ (excluding } H_0). \tag{3.29}$$

For this the LMMP test is based on critical regions of the form (3.20), i.e.

$$d^* = Y'AY \leq c_3, \tag{3.30}$$

---

[3] Similar format of ρ_1 and ρ_2 in relation to the 3SLR model is defined by (2.6).

where c_3 is a suitably chosen constant and in this case $A = A_1 + A_2$, in which

$$A_1 = \bigoplus_{i=1}^{m} A_{1i}$$

$$A_2 = \bigoplus_{i=1}^{m} A_{2i}$$

and by (3.27)

$$A_{1i} = I_{N_i} \otimes I_T - e_{N_i} e'_{N_i} \otimes e_T e'_T = \left(I_{N_i T} - E_{N_i} \otimes E_T \right),$$

$$A_{2i} = I_{N_i} \otimes I_T - I_{N_i} \otimes e_T e'_T = \left(I_{N_i T} - I_{N_i} \otimes E_T \right).$$

The rest of this section is constrained to the case of equal clusters, i.e., $N_i = s$, and also of equal subclusters, i.e., $m(i,j) = T$. Therefore d^* in (3.30), can be expressed as

$$d^* = Y' A_1 Y + Y' A_2 Y$$

$$= Y' \left(I_n - D_1 \right) Y + Y' \left(I_n - D_2 \right) Y$$

$$= 2Y'Y - \left(Y' D_1 Y + Y' D_2 Y \right) = Y(2I_n - D_1 - D_2)Y = Y'AY$$

$$= 2 \sum_{i=1}^{m} \sum_{j=1}^{s} \sum_{k=1}^{T} y_{ijk}^2 - \left[\sum_{i=1}^{m} \left(\sum_{j=1}^{s} \sum_{k=1}^{T} y_{ijk} \right)^2 + \sum_{i=1}^{m} \sum_{j=1}^{s} \left(\sum_{k=1}^{T} y_{ijk} \right)^2 \right] \quad (3.31)$$

where,

$$A = \left(2I_n - D_1 - D_2 \right)$$

$$A_1 = I_n - D_1,$$

$$A_2 = I_n - D_2,$$

$$D_1 = I_m \otimes E_{sT},$$

$$D_2 = I_{ms} \otimes E_T,$$

in which E_T and E_{sT} are $T \times T$ and $sT \times sT$ matrices of ones. Note that d^* in (3.30) or in (3.31) is a quadratic form in normal variables, and therefore its critical value, c_3, may be obtained by evaluating the probabilities of the form

$$\Pr\left[d^* < c_3\right] = \Pr\left[2Y'Y - Y'\left(D_1 + D_2\right)Y < c_3\right]$$

$$= \Pr\left[Y'\left\{2I_n - \left(D_1 + D_2\right)\right\}Y < c_3\right]$$

$$= \Pr\left[\sum_{i=1}^{n} \lambda_i \xi_i^2 < c_3\right] \tag{3.32}$$

where c_3 is a suitably chosen constant, $\xi = \left(\xi_1, \ldots, \xi_n\right)' \sim N\left(0, I_n\right)$, and λ_i for $i=1,\ldots,n$ are the eigenvalues of the matrix $\{2I_n - (D_1 + D_2)\}$, namely 2, $(2-T)$ and $\{2-T(s+1)\}$ with multiplicities of $ms(T-1)$, $m(s-1)$ and m respectively. Therefore (3.32) can be written as

$$\Pr\left(d^* < c_3\right) = \Pr\left[2\chi^2_{ms(T-1)} + \left(2-T\right)\chi^2_{m(s-1)} + \left(2-T(s+1)\right)\chi^2_m < c_3\right].$$

The next section will explain how the PO test for testing (3.28) against (3.29) can be constructed, particularly when dealing with the three-stage SSMC distribution model.

Point Optimal Test

For the same testing problem discussed in the previous section, a PO test is constructed, which would be used to obtain the maximum attainable power, i.e. the PE over the alternative hypothesis parameter space. For any particular test its performance can be assessed by checking how close its power is to the PE. In this section the PO test statistic for testing $H_0: \rho_1 = \rho_2 = 0$ against the $H_a: \rho_1 > 0, \rho_2 > 0$ is introduced.

Based on (3.25), $(\rho_1, \rho_2)' = (\rho_{11}, \rho_{22})'$; a known point at which optimal power in the alternative parameter space is obtained. By following the procedure of the BO test, with error covariance matrix (3.26), one can obtain the PO test for testing simple null hypothesis, $H_0: \rho_1 = \rho_2 = 0$ against a simple alternative hypothesis $H_a : \rho_1 = \rho_{11} > 0, \rho_2 = \rho_{22} > 0,$ that is to reject the null hypothesis for small values of a test statistic

$$r\left(\rho_{11}, \rho_{22}\right) = Y'\left(\Omega^{-1}\left(\rho_{11}, \rho_{22}\right) - I_n\right)Y,$$

where

$$\Omega^{-1}\left(\rho_{11}, \rho_{22}\right) = \left[\left(1 - \rho_{11} - \rho_{22}\right)I_n + \rho_{11}D_1 + \rho_{22}D_2\right]^{-1}.$$

The above inverse matrix can be evaluated by using Searle and Henderson's (1979) procedure, in which $\Omega^{-1}(\rho_{11}, \rho_{21})$ is obtained as follows

$$\Omega^{-1}\left(\rho_{11}, \rho_{22}\right) = \left(1 - \rho_{11} - \rho_{22}\right)^{-1}\left[I_n - \left(1 - \rho_{11} - \rho_{22} + T\rho_{22}\right)^{-1}\right.$$

$$\left\{\rho_{11}\left(1 - \rho_{11} - \rho_{22}\right)\left(1 - \rho_{11} + T\rho_{22} + sT\rho_{11}\right)^{-1}D_1 + \rho_{22}D_2\right\} \quad (3.33)$$

where D_1 and D_2 are defined earlier by (3.31), i.e. $D_1 = \left(I_m \otimes E_{sT}\right)$ and

$$D_2 = \left(I_{ms} \otimes E_T\right).$$

Note that $r(\rho_{11}, \rho_{22})$ is a quadratic form in normal variables, and therefore its critical value, c_α, may be obtained by evaluating the probabilities of the form

$$\Pr\left[r\left(\rho_{11}, \rho_{22}\right) < c_\alpha \right]$$

$$= \Pr\left[Y'\left(\Omega^{-1}\left(\rho_{11}, \rho_{22}\right) - I_n\right)Y < c_\alpha \right]$$

$$= \Pr\left[\sum_{i=1}^{n} \lambda_i \xi_i^2 < c_\alpha \right]$$

$$= \Pr\left[\sum_{i=1}^{ms(T-1)} \lambda_i \xi_i^2 + \sum_{i=ms(T-1)+1}^{m(s-1)} \lambda_i \xi_i^2 + \sum_{i=m(s-1)+1}^{m} \lambda_i \xi_i^2 < c_\alpha \right]$$

$$= \Pr\left[\left\{ \left(1 - \rho_{11} - \rho_{22}\right)^{-1} - 1 \right\} \chi_{ms(T-1)}^2 + \left\{ \left(1 - \rho_{11} - \rho_{22}\left(1 - T\right)\right)^{-1} - 1 \right\} \right.$$

$$\left. \chi_{m(s-1)}^2 + \left\{ \left(1 + \rho_{11}\left(sT - 1\right) + \rho_{22}\left(T - 1\right)\right)^{-1} - 1 \right\} \chi_m^2 < c_\alpha \right] \qquad (3.34)$$

where $\xi = (\xi_1,...,\xi_n)' \sim IN(0, I_n)$ and λ_i, $i=1,...,n$ are the eigenvalues of the matrix $(\Omega^{-1}(\rho_{11}, \rho_{22}) - I_n)$, namely $\{(1-\rho_{11}-\rho_{22})^{-1}-1\}$, $\{(1-\rho_{11}-\rho_{22}(1-T)^{-1})-1\}$ and $\{(1+\rho_{11}(sT-1) + \rho_{22}(T-1))^{-1}-1\}$ with multiplicities of $ms(T-1)$, $m(s-1)$ and m, respectively. In order to find c_α such that (3.34) equals the required significance level, α, (3.34) can be evaluated by standard numerical algorithms, which were mentioned in earlier sections.

To calculate the power of the $r(\rho_{11}, \rho_{22})$ test, there is a need to decompose the error covariance matrix $\Omega(\rho_1, \rho_2)$ that is

$$\Omega\left(\rho_1, \rho_2\right) = \Omega^{1/2}\left(\rho_1, \rho_2\right)\left[\Omega^{1/2}\left(\rho_1, \rho_2\right)\right]',$$

where

$$\Omega^{1/2}\left(\rho_1,\rho_2\right) = \overset{m}{\underset{i=1}{\oplus}}\Omega_i^{1/2}\left(\rho_1,\rho_2\right),$$

such that the elements of the diagonal components matrix, $\Omega^{1/2}(\rho_1,\rho_2)$ are given in Wu and Bhatti (1994) appendix, and the formula for the $\Omega_i^{1/2}(\rho_1,\rho_2)$ is expressed below

$$\Omega_i^{1/2}\left(\rho_1,\rho_2\right) = \sqrt{\lambda_1}\,\frac{e_{N_iT}e'_{N_iT}}{N_iT} + \sqrt{\lambda_2}\left(I_{N_i}\otimes I_T - I_{N_i}\otimes\frac{e_T e'_T}{T}\right)$$

$$+ \sqrt{\lambda_3}\left(I_{N_i}\otimes\frac{e_T e'_T}{T} - \frac{e_{N_iT}e'_{N_iT}}{N_iT}\right)$$

where

$$\lambda_1 = 1 - \rho_1 - \rho_2 + N_iT\rho_1 + T\rho_2,$$

$$\lambda_2 = 1 - \rho_1 - \rho_2,$$

$$\lambda_3 = 1 - \rho_1 - \rho_2 + T\rho_2. \tag{3.35}$$

For any critical value c_α, the power of the critical region $r\left(\rho_{11},\rho_{22}\right) < c_\alpha$ is obtained by evaluating the probabilities of the form

$$\Pr\left[r\left(\rho_{11},\rho_{22}\right) < c_\alpha\right]$$

$$= \Pr\left[Y'\left(\Omega^{-1}\left(\rho_{11},\rho_{22}\right) - I_n\right)Y < c_\alpha \middle| Y \sim N\left(0,\Omega\left(\rho_1,\rho_2\right)\right)\right]$$

$$= \Pr\left[z'\left(\Omega^{1/2}\right)'\left(\Omega^{-1}\left(\rho_{11},\rho_{22}\right) - I_n\right)\Omega^{1/2}z < c_\alpha \middle| z \sim N\left(0,I_n\right)\right]$$

$$= \Pr\left[\sum_{i=1}^{n} \lambda_i \xi_i^2 < c_\alpha \right]$$

$$= \Pr\left[a\chi_{ms(T-1)}^2 + b\chi_{m(s-1)}^2 + c\chi_m^2 < c_\alpha \right], \tag{3.36}$$

where $\xi_i \sim IN(0,1)$, χ_j^2 denotes the usual chi-square distribution with j degrees of freedom and λ_i's are the eigenvalues of

$$\left(\Omega^{1/2} \right)' \left(\Omega^{-1}(\rho_{11}, \rho_{22}) - I_n \right) \Omega^{1/2},$$

or equivalently, the eigenvalues of

$$\left(\Omega^{-1}(\rho_{11}, \rho_{22}) - I_n \right) \Omega(\rho_1, \rho_2)$$

$$= \Omega^{-1}(\rho_{11}, \rho_{22}) \Omega(\rho_1, \rho_2) - \Omega(\rho_1, \rho_2),$$

where

$$\Omega^{1/2} = \Omega^{1/2}(\rho_1, \rho_2).$$

It follows from lemma two together with lemma three of appendix 3A that the eigenvalues of the latter matrix are

$$a = \frac{\left(1 - \rho_1 - \rho_2\right)}{\left(1 - \rho_{11} - \rho_{22}\right)} \left[\rho_{11} + \rho_{22}\right] \tag{3.37}$$

$$b = \frac{\left(1 - \rho_1 - \rho_2(1-T)\right)}{\left(1 - \rho_{11} - \rho_{22}(1-T)\right)} \left\{ \rho_{11} + \rho_{22}(1-T) \right\}, \tag{3.38}$$

and

$$c = \frac{\left(1 - \rho_1(1-sT) - \rho_2(1-T)\right)}{\left(1 - \rho_{11}(1-sT) - \rho_{22}(1-T)\right)}\left\{\rho_{11}(1-sT) + \rho_{22}(1-T)\right\}, \quad (3.39)$$

with multiplicities of $ms(T-1)$, $m(s-1)$ and m.

In the special cases in which the power is being calculated at a given point

$$(\rho_1, \rho_2)' = (\rho_{11}, \rho_{22})',$$

then (3.37), (3.38) and (3.39) reduce to

$$\left(\rho_{11} + \rho_{22}\right), \left(\rho_{11} + \rho_{22}(1-T)\right) \text{ and } \left\{\rho_{11}(1-sT) + (1-T)\right\},$$

so that the probably expression of (3.36) can be written as

$$\Pr\left[\chi^2_{ms(T-1)} + \left(1 - \frac{T\rho_{22}}{\rho_{11} + \rho_{22}}\right)\chi^2_{m(s-1)} + \left(1 - \frac{T\left(s\rho_{11} + \rho_{22}\right)}{\rho_{11} + \rho_{22}}\right)\chi^2_m < \frac{c_\alpha}{\rho_{11} + \rho_{22}} \right]. \quad (3.40)$$

Under these special circumstances, the power of the $r(\rho_1, \rho_2)$ test can be computed by solving (3.40). In the next section the PO test for testing subcluster effects in the presence of main cluster effects will be constructed.

3.6. PO TEST FOR TESTING SUBCLUSTER EFFECTS

In the previous section the PO test for testing whether intracluster and intrasubcluster correlation coefficients are equal to zero, against the alternative that they have some positive value derived. The main focus of this section is on testing the intrasubcluster correlation coefficient in the presence of intracluster correlation. Detailed applications of such a testing problem in the case of 3SLR model are given in chapter six of this book.

In this section our objective is to test

$$H_0 : \rho_2 = 0, \rho_1 > 0$$

$$(3.41)$$

against the one-sided alternative

$$H_a : \rho_2 > 0, \rho_1 > 0.$$ $$(3.42)$$

This problem is more complex than the testing problem of the PO test in the previous section. In this case both the null and the alternative hypotheses are composite. However, under H_0 the density function of Y is

$$f_0\left(Y:\rho_1,0\right) = \left(2\pi\right)^{-n/2} \left|\Omega_0\right|^{-1/2} \exp\left[-\frac{1}{2}Y'\Omega_0^{-1}Y\right]$$

while under H_a it is given by

$$f_a\left(Y:\rho_1,\rho_2\right) = \left(2\pi\right)^{-n/2} \left|\Omega_1\right|^{-1/2} \exp\left[-\frac{1}{2}Y'\Omega_1^{-1}Y\right]$$

where $\Omega_0 = \Omega\left(\rho_1,0\right)$ and $\Omega_1 = \Omega\left(\rho_1,\rho_2\right)$.

Note that $\Omega(\rho_1,\rho_2)$ is defined by (3.26), where $0 \le \rho_1 + \rho_2 \le 1$ and ρ_1 and ρ_2 are unknown parameters. For the purpose of test construction, there are a couple of options for deciding on the range of ρ_1 values. The first option is $0 \le \rho_1 \le 1$, which seems unlikely because $0 \le \rho_1 + \rho_2 \le 1$. The other option, that might be reasonable to assume is $0 \le \rho_1 \le 0.5$. Before this it is important to consider the simple problem of testing,

$$H_0'' : \left(\rho_1,\rho_2\right)' = \left(\rho_{10},0\right)',$$

against the simple alternative hypothesis

$$H_a^{''} : \left(\rho_1, \rho_2 \right) = \left(\rho_{11}, \rho_{21} \right),$$

where $0 \le \rho_{10} \le 0.5$, $0 \le \rho_{11} \le 0.5$ and $0 \le \rho_{21} \le 0.5$ are known fixed values. The Neyman-Pearson Lemma implies that a most powerful test can be based on the critical region of the form

$$r \left(\rho_{10}, \rho_{11}, \rho_{21} \right) = Y' \left(\Omega_{11}^{-1} - \Omega_{10}^{-1} \right) Y < c_\alpha , \tag{3.43}$$

where c_α is an appropriate critical value.

Note that $r(\rho_{10}, \rho_{11}, \rho_{21})$ in (3.43) is a quadratic form in normal variables, and therefore its critical values, c_α, for the required level of significance, α, can be obtained by evaluating the probabilities of the form

$$= \Pr \left[r \left(\rho_{10}, \rho_{11}, \rho_{21} \right) < c_\alpha \middle| Y \sim N \left(0, \Omega \left(\rho_{10}, 0 \right) \right) \right]$$

$$= \Pr \left[Y' \left(\Omega_{11}^{-1} - \Omega_{10}^{-1} \right) Y < c_\alpha \middle| Y \sim N \left(0, \Omega_0 \right) \right]$$

$$= \Pr \left[\left(\Omega_0^{-1/2} Y \right)' \left(\Omega_0^{1/2} \right)' \left(\Omega_{11}^{-1} - \Omega_{10}^{-1} \right) \Omega_0^{-1/2} \left(\Omega_0^{-1/2} Y \right) < c_\alpha \middle| Y \sim N \left(0, \Omega_0 \right) \right]$$

$$= \Pr \left[z' \left(\Omega_0^{1/2} \right)' \left(\Omega_{11}^{-1} - \Omega_{10}^{-1} \right) \Omega_0^{1/2} z < c_\alpha \middle| z \sim N \left(0, I_n \right) \right]$$

$$= \Pr \left[\sum_{i=1}^{m} \lambda_i \xi_i^2 < c_\alpha \right], \tag{3.44}$$

Where $\xi_i \sim IN \left(0, 1 \right)$, and λ_i's are the eigenvalues of $\left[\left(\Omega_0^{1/2} \right)' \left(\Omega_{11}^{-1} - \Omega_{10}^{-1} \right) \Omega_0^{1/2} \right]$ or equivalently, the eigenvalues of

$$\left[\left(\Omega_{11}^{-1} - \Omega_{10}^{-1} \right) \Omega_0 \right]$$

(3.45)

in which $\Omega_{11} = \Omega\left(\rho_{11}, \rho_{22}\right), \Omega_{10} = \Omega\left(\rho_{10}, 0\right)$ and $\Omega_0 = \Omega\left(\rho_1, 0\right)$.

Using lemmas three and four of appendix 3A, together with lemma two of the same appendix, the eigenvalues of the latter matrix can be found, as

$$a_1 = \left(1 - \rho_1\right)\left\{ \frac{1}{1 - \rho_{11} - \rho_{22}} - \frac{1}{1 - \rho_{10}} \right\},$$

(3.46)

$$b_1 = \frac{\left(1 - \rho_1\right)}{1 - \rho_{11} - \rho_{22}}\left\{ \frac{1 - \rho_{11} - \rho_{22}}{1 - \rho_{11} - \rho_{22} + T\rho_{22}} - \frac{1}{1 - \rho_{10}} \right\},$$

(3.47)

and

$$c_1 = \left(1 - \rho_1 + sT\rho_1\right)\left[\frac{1}{\left(1 - \rho_{11} - \rho_{22} + T\rho_{22} + sT\rho_{11}\right)} \right.$$

$$\left. - \left(\frac{1}{\left(1 - \rho_{10}\right)\left(1 - \rho_1 + sT\rho_1\right)} - \frac{\rho_{10}}{\left(1 - \rho_{10}\right)\left(1 - \rho_{10} + sT\rho_{10}\right)} \right) \right],$$

(3.48)

with multiplicities $ms(T-1)$, $m(s-1)$ and m.

Therefore, one can write (3.44) critical region expression in the χ^2 variates as

$$\Pr\left[a_1 \chi_{ms(T-1)}^2 + b_1 \chi_{m(s-1)}^2 + c_1 \chi_m^2 < c_\alpha \right],$$

(3.49)

where a_1, b_1 and c_1 are given by (3.46), (3.47) and (3.48). Thus a test based on $r\left(\rho_{10}, \rho_{11}, \rho_{21}\right)$ in (3.43), is most powerful in the neighbourhood of $\left(\rho_1, \rho_2\right) = \left(\rho_{11}, \rho_{21}\right)$, and therefore it is a PO test of a simple null H_0''

against a simple alternative $H_a^{''}$. For the wider problem of testing (3.41) against (3.42), it may not necessarily be PO, as for this testing problem the critical value should be found by solving

$$\sup_{0<\rho_1\leq0.5} \Pr\left[r\left(\rho_{10},\rho_{11},\rho_{21}\right)<c_\alpha^*\Big|y\sim N\left(0,\Omega\left(\rho_1,0\right)\right)\right]=\alpha \quad (3.50)$$

for c_α^*.

In general, $c_\alpha < c_\alpha^*$ so the two critical regions are different. As the range of ρ_1 is closed

$$\Pr\left[r\left(\rho_{10},\rho_{11},\rho_{21}\right)<c_\alpha^*\Big|y\sim N\left(0,\Omega\left(\rho_1,0\right)\right)\right]=\alpha$$

is true for at least one ρ_1 value. If one can find such a ρ_{10} to let $c_\alpha = c_\alpha^*$ can be found, then the test based on $r\left(\rho_{10},\rho_{11},\rho_{21}\right)$ is MP and is a PO test of (3.41) against (3.42). However, in general such a ρ_{10} value may not exist, or perhaps may exist for some combinations of ρ_{11}, ρ_{21}, m, s and T values, but not for others. If the value of the ρ_{10} does exist (as will be seen in chapter six dealing with the 3SLR model), then (3.46), (3.47) and (3.48) become

$$a_1' = \left\{\frac{1-\rho_{10}}{1-\rho_{11}-\rho_{22}}-1\right\},$$

$$b_1' = \frac{1}{1-\rho_{11}-\rho_{22}}\left\{\frac{\left(1-\rho_{11}-\rho_{22}\right)\left(1-\rho_{10}\right)}{\left(1-\rho_{11}-\rho_{22}+T\rho_{22}\right)}-1\right\}$$

and

$$c_1' = \left\{\frac{1-\rho_{10}+sT\rho_{10}}{\left(1-\rho_{11}-\rho_{22}+T\rho_{22}+sT\rho_{11}\right)}-1\right\},$$

Therefore c_α can be obtained by solving the left hand side of

$$\Pr\left[a_1' \chi^2_{ms(T-1)} + b_1' \chi^2_{m(s-1)} + c_1' \chi^2_m < c_\alpha \right] = \alpha,$$

where α is the desired level of significance, and χ^2_j is the usual chi-squared distribution with j degrees of freedom. If in case ρ_{10} does not exist, then one can look for an approximate PO test, which was discussed in chapter two.

3.7. CONCLUDING REMARKS

This chapter considered a nx1 dimensional random vector, Y, which follows either the two- or the three-stage SSMC distribution, and derived some optimal tests for detecting the intra cluster correlation coefficient between observations within clusters and subclusters. This chapter also considered more complicated problems for detecting subcluster correlation in the presence of main cluster correlation.

In the case of the two-stage SSMC distribution model, the PE with the powers of SenGupta's LB test and two versions of the BO tests (ie, r0.5 and r0.8) at the five percent level of significance for different clusters and sample sizes, were computed and compared. Selected one and five percent significance points, c_α and their associated ρ_1 values are given in tables 3.1 and 3.2 of appendix 3C whereas the power of comparison for the various values of k and m are presented in tables 3.3, 3.4 and 3.5 of the appendix. The main findings of this chapter are that the BO tests are usually superior to the LB test, sometimes by a wide margin, whilst the power differences between the PE and BO tests are very small, with the biggest difference being 0.031. On the basis of these results it is argued that the BO tests are approximately UMP tests, at least at the five percent level of significance.

Encouraged by the small sample power performances of BO tests for the two-stage SSMC model, this chapter enabled us to derive POI and LMMPI tests for the three-stage SSMC model. The next logical step is to consider a model based on the SMC distribution (ie, when σ^2 is unknown) and discover the consequences of applying optimal tests to it. This is done in the next chapter.

APPENDIX 3A. SOME USEFUL MATRIX RESULTS

This appendix gives some important matrix results which are useful in chapters three, four, five and six.

Lemma 1:

Let $\Delta(\rho) = (1-\rho)I_n + \rho.D$, where $\dfrac{-1}{n-1} < \rho < 1$, D is given by (3.5) and

$n = mk$, then one can have the following expressions.

i. $\left|\Delta(\rho)\right| = \left[(1-\rho)^{k-1}\left[1+(k-1)\right]\rho\right]^m$.

ii. $\Delta^{-1}(\rho) = (1-\rho)^{-1}\left[I_n - \rho\left\{1+(k-1)\rho\right\}^{-1}D\right]$.

iii. The eigenvalues of $\Delta^{-1}(\rho)$ are $\dfrac{1}{(1-\rho)}$ and $\dfrac{1}{\left[1+(k-1)\rho\right]}$ with multiplicity $m(k-1)$ and m, respectively.

iv. The eigenvalues of $\Delta(\rho)$ are $(1-\rho)$ and $\left[1+(k-1)\rho\right]$ with multiplicity $m(k-1)$ and m, respectively.

v. The matrix $\Delta(\rho)$, can be expressed in functional form as

$$\Delta(\rho) = \left(f_1(\rho)I_n + f_2(\rho)D\right)^{-1},$$

where

$$f_1(\rho) = (1-\rho)^{-1}, f_2(\rho) = -\rho\left[(1-\rho)\left\{1+(k-1)\rho\right\}\right]^{-1}.$$

The proof is straightforward and is therefore omitted. For the special case when $m=1$, this lemma reduces to Rao (1973, p. 67, problem 1.1)

Lemma 2

If A is nxn and B is nxn, then the eigenvalues of AB are the engenvalues of BA, although the eigenvectors of the two matrices need not be the same (see Graybill, 1969, p. 208).

Lemma 3

Let $D_1 = I_m \otimes E_{sT}, D_2 = I_{ms} \otimes E_T, n = msT$ and $\lambda_i(A)$ denote the ith eigenvalue of the matrix A, then

$$\lambda_i(D_1 + D_2) = \lambda_i(D_1) + \lambda_i(D_2),$$

for $i=1,...,n$. These eigenvalues are $T(s+1)$, T and zeroes with multiplicities of m, $m(s-1)$, and $ms(T-1)$, respectively.

Lemma 4

Let

$$\Omega(\rho_1, \rho_2) = (1 - \rho_1 - \rho_2)I_n + \rho_1 D_2 + \rho_2 D_2, \text{ where } 0 \le \rho_1 + \rho_2 \le 1, \text{ then}$$

i.   The eigenvalues of $\Omega(\rho_1, \rho_2)$ are $(1 - \rho_1 - \rho_2)$,

$$\{1 - \rho_1 - \rho_2(1-T)\} \text{ and } \{1 + p_1(sT-1) + \rho_2(T-1)\}$$

with multiplicities $ms(T-1)$, $m(s-1)$, and m, respectively.

ii. $\Omega^{-1}(\rho_1, \rho_2) = (1 - \rho_1 - \rho_2)^{-1}\left[I_n - (1 - \rho_1 - \rho_2)(T+1)^{-1}\right]$

$$\left\{\rho_1(1 - \rho_1 - \rho_2)(1 - \rho_1 - \rho_2 + T(\rho_2 + s\rho_1))^{-1} D_1 + \rho_2 D_2\right\}$$

iii.   The eigenvalues of $\Omega^{-1}\left(\rho_1,\rho_2\right)$ are the reciprocal of the eigenvalues of $\Omega\left(\rho_1,\rho_2\right)$

The results of lemmas three and four are the special cases which are obtained by using Searle and Henderson's (1979) general procedure.

APPENDIX 3B: TABULATION OF RESULTS OF THE COMPARATIVE STUDY

Table 3.1. Selected values of ρ_1 and c_α for the $r_{0.5}$ test at the one and five percent significance levels

K	α	m=5		m=10		m=15		m=25	
		ρ_1	c_α	ρ_1	c_α	ρ_1	c_α	ρ_1	c_α
2	0.01	.6694	10^{-4}	.6581	10^{-5}	.5577	10^{-5}	.4451	10^{-5}
	0.05	.8329	-10^{-5}	.4973	-10^{-5}	.4124	-10^{-6}	.3233	10^{-5}
3	0.01	.6269	.3779	.4531	.2885	.3700	.2394	.2850	.1867
	0.05	.4610	.2778	.3227	.2055	.2012	.1690	.2002	.1311
4	0.01	.4996	.6146	.3461	.4448	.2779	.3618	.2107	.2770
	0.05	.3530	.4342	.2403	.3088	.1924	.2504	.1459	.1918
5	0.01	.4151	.7748	.2803	.5430	.2228	.4364	.1674	.3308
	0.05	.2864	.5346	.1917	.3714	.1525	.2988	.1149	.2271
6	0.01	.3550	.8900	.2356	.6105	.1860	.4869	.1389	.3664
	0.05	.2412	.6046	.1596	.4136	.1264	.3309	.0949	.2503
7	0.01	.3102	.9769	.2032	.6598	.1597	.5233	.1187	.3918
	0.05	.2084	.6562	.1367	.4440	.1079	.3538	.0808	.2667
8	0.01	.2754	1.045	.1787	.6973	.1399	.5507	.1037	.4109
	0.05	.1835	.6960	.1196	.4670	.0942	.3709	.0704	.2792
9	0.01	.2477	1.099	.1594	.7270	.1245	.5722	.0920	.4257
	0.05	.1639	.7275	.1063	.4849	.0836	.3843	.0623	.2884
10	0.01	.2250	1.442	.1439	.7509	.1121	.5895	.0828	.4378
	0.05	.1481	.7531	.0957	.493	.0751	.3949	.0560	.2959

Table 3.2. Selected values of ρ_1 and c_α for the $r_{0.8}$ test at the one and five percent significance levels

K	α	m=5		m=10		m=15		m=25	
		ρ_1	c_α	ρ_1	c_α	ρ_1	c_α	ρ_1	c_α
2	0.01	.9114	3.068	.7767	3.906	.6835	4.289	.5656	4.653
	0.05	.8189	2.756	.6624	3.331	.5697	3.575	.4615	3.796
3	0.01	.7657	4.680	.5942	5.338	.4980	5.546	.3910	5.672
	0.05	.6469	3.953	.4821	4.331	.3979	4.431	.3086	4.476
4	0.01	.6584	5.784	.4833	6.212	.3940	6.263	.3008	6.211
	0.05	.5384	4.730	.3823	4.914	.3082	4.900	.2336	4.824
5	0.01	.5788	6.621	.4084	6.819	.3268	6.739	.2448	6.552
	0.05	.4633	5.299	.3178	5.306	.2523	5.203	.1883	5.039
6	0.01	.5173	7.289	.3541	7.270	.2794	7.080	.2066	6.788
	0.05	.4077	5.744	.2724	5.592	.2138	5.416	.1579	5.187
	0.01	.4684	7.840	.3128	7.621	.2442	7.337	.1788	6.962
	0.05	.3646	6.103	.2385	5.811	.1856	5.576	.1360	5.294
8	0.01	.4284	8.305	.2803	7.901	.2170	7.539	.1577	7.095
	0.05	.3302	6.401	.2123	5.983	.1640	5.699	.1195	5.376
9	0.01	.3950	8.705	.2540	8.132	.1952	7.701	.1410	7.201
	0.05	.3019	6.654	.1913	6.124	.1470	5.797	.1065	5.441
10	0.01	.3666	9.052	.2323	8.324	.1775	7.835	.1275	7.287
	0.05	.2783	6.871	.1741	6.240	.1331	5.878	.0961	5.493

Table 3.3. Comparison of the PE with the powers of LB, $r_{0.5}$ and $r_{0.8}$ tests at the five percent significance level for $m=10$

Tests	$\rho=.05$	.1	.2	.3	.4	.5	.6	.8
				k=2				
PE	.069	.093	.155	.242	.357	.504	.683	.983
LB	.069	.092	.153	.213	.325	.430	.538	.737
$r_{0.5}$	.067	.088	.147	.234	.354	.504	.675	.952
$r_{0.8}$	.065	.083	.135	.215	.331	.489	.680	.978
				k=3				
PE	.088	.139	.279	.456	.650	.829	.951	1.000
LB	.088	.139	.270	.423	.572	.701	.802	.926
$r_{0.5}$	.085	.135	.275	.456	.646	.809	.920	1.000
$r_{0.8}$	.081	.125	.255	.438	.646	.829	.946	1.000
				k=4				
PE	.109	.192	.406	.636	.830	.950	.994	1.000
LB	.109	.190	.389	.582	.735	.841	.909	.974
$r_{0.5}$	.106	.188	.405	.633	.812	.922	.975	.999
$r_{0.8}$	.100	.174	.387	.631	.830	.945	.990	1.000
				k=6				
PE	.155	.305	.622	.851	.964	.997	1.000	1.000
LB	.155	.301	.589	.785	.893	.949	.976	.995
$r_{0.5}$	.153	.303	.621	.836	.941	.983	.996	1.000
$r_{0.8}$	.144	.288	.617	.851	.958	.992	.999	1.000
				k=10				
PE	.261	.521	.861	.977	.999	1.000	1.000	1.000
LB	.259	.509	.820	.936	.977	.991	.997	.999
$r_{0.5}$	.259	.521	.849	.960	.990	.998	1.000	1.000
$r_{0.8}$	.249	.513	.860	.971	.996	1.000	1.000	1.000

Table 3.4. Comparison of the PE with the powers of LB, $r_{0.5}$ and $r_{0.8}$ tests at the five percent significance level for m=15

Tests	ρ=.05	.1	.2	.3	.4	.5	.6	.8
				k=2				
PE	.074	.105	.191	.315	.478	.668	.852	.999
LB	.074	.104	.188	.299	.431	.569	.699	.889
$r_{0.5}$	.072	.101	.186	.312	.478	.663	.833	.993
$r_{0.8}$	.070	.096	.174	.295	.464	.665	.851	.998
				k=3				
PE	.098	.167	.359	.591	.804	.943	.994	1.000
LB	.098	.165	.346	.547	.720	.844	.922	.985
$r_{0.5}$	.096	.163	.358	.590	.792	.921	.979	1.000
$r_{0.8}$	.092	.154	.343	.585	.804	.939	.990	1.000
				k=4				
PE	.125	.237	.522	.783	.939	.992	1.000	1.000
LB	.125	.235	.499	.724	.867	.942	.977	.997
$r_{0.5}$	.123	.235	.522	.774	.920	.979	.996	1.000
$r_{0.8}$	.118	.224	.514	.783	.935	.988	.999	1.000
				k=6				
PE	.186	.388	.763	.947	.995	1.000	1.000	1.000
LB	.186	.381	.726	.901	.967	.990	.997	1.000
$r_{0.5}$	.185	.388	.757	.932	.986	.998	1.000	1.000
$r_{0.8}$	.178	.378	.763	.944	.992	.999	1.000	1.000
				k=10				
PE	.327	.650	.950	.997	1.000	1.000	1.000	1.000
LB	.325	.635	.922	.985	.997	.999	1.000	1.000
$r_{0.5}$	.326	.649	.940	.992	.999	1.000	1.000	1.000
$r_{0.8}$	.320	.649	.947	.995	1.000	1.000	1.000	1.000

Table 3.5. Comparison of the PE with the powers of LB, $r_{0.5}$ and $r_{0.8}$ tests at the five percent significance level for $m=25$

Tests	$\rho=.05$	.1	.2	.3	.4	.5	.6	.8
				$k=2$				
PE	.082	.126	.259	.449	.672	.868	.975	1.000
LB	.082	.126	.253	.424	.609	.773	.889	.986
$r_{0.5}$	.080	.123	.256	.449	.669	.854	.960	1.000
$r_{0.8}$	.078	.119	.246	.440	.670	.867	.972	1.000
				$k=3$				
PE	.115	.216	.499	.780	.946	.995	1.000	1.000
LB	.115	.215	.480	.731	.889	.964	.990	1.000
$r_{0.5}$	.114	.214	.499	.774	.932	.988	.999	1.000
$r_{0.8}$	.110	.207	.493	.780	.943	.993	1.000	1.000
				$k=4$				
PE	.154	.320	.700	.930	.994	1.000	1.000	1.000
LB	.153	.316	.670	.887	.970	.994	.999	1.000
$r_{0.5}$	.152	.319	.697	.919	.987	.999	1.000	1.000
$r_{0.8}$	.148	.311	.699	.928	.991	.999	1.000	1.000
				$k=6$				
PE	.243	.529	.912	.994	1.000	1.000	1.000	1.000
LB	.242	.519	.884	.981	.997	1.000.	1.000	1.000
$r_{0.5}$	.242	.529	.905	.989	.999	1.000	1.000	1.000
$r_{0.8}$	.237	.526	.911	.992	1.000	1.000	1.000	1.000
				$k=10$				
PE	.443	.820	.994	1.000	1.000	1.000	1.000	1.000
LB	.440	.805	.987	.999	1.000	1.000	1.000	1.000
$r_{0.5}$	.443	.817	.991	1.000	1.000	1.000	1.000	1.000
$r_{0.8}$	.440	.820	.993	1.000	1.000	1.000	1.000	1.000

4. TESTS FOR CLUSTER EFFECTS OF THE SMC MODEL

4.1. INTRODUCTION

In chapter three, it was noted that the LB, BO, PO and LMMP tests have certain small sample optimal power properties for two- and three-stage SSMC distribution models. The objective of this chapter is to consider analogous tests for the case of SMC distribution models, in which σ^2 is an unknown scalar parameter. This chapter genralizes the results of the previous chapter in constructing the PO invariant (POI) and LMMP invariant (LMMPI) test and demonstrates that the POI test for the two-stage SMC model is UMP invariant (UMPI). The most attractive feature of the POI test in such a case is that its power and critical values are obtainable from the standard F distribution. Among other things this chapter also explores the possibility of developing some optimal tests based on the three-stage SMC distribution model.

The structure of this chapter is as follows. In the next section, the two-stage SMC model is introduced and the POI test is constructed, which illustrates how to calculate the power and the critical values of the POI test with the help of the standard F distribution. Also, the next section provides proof of the fact that the test is UMPI, and constructs a LMMPI test for a more complicated model, where means are known but variances and the equicorrelation coefficient are unknown. In addition to this, the following section demonstrates how to derive LMMPI and POI tests for the three-stage SMC model, and develops the POI test in testing subcluster effects in the presence of main cluster effects, whilst the final section of this chapter contains some concluding remarks.

4.2. TWO-STAGE SMC MODEL AND THE TESTS

In chapter three so far testing problems associated with the SSMC model have been considered. Although the definition and examples of the model which follow the SMC distribution has been discussed in chapter two, no test for testing equicorrelation coefficient based on SMC model was developed in the process. The objective of this section is to construct some optimal tests based on the two-stage SMC model. The model under consideration is

$$Y \sim N\left(0, \sigma^2 \Delta(\rho)\right), \tag{4.1}$$

where the details of the matrix $\Delta(\rho)$ is given by (3.4) via (3.1) or equivalently (2.3). In addition to this, throughout this chapter a more practical assumption that σ^2 is an unknown scalar parameter is made. Based on the model (4.1), one can easily construct the PO test, which is given in the next sub-section and called a POI test.

Point Optimal Invariant Test

Following the previous chapter, the problem of interest in this subsection is to test

$$H_0 : \rho = 0$$

against

$$H_a : \rho > 0.$$

It is important to note that this testing problem is invariant (in the sense of Lehmann, 1959 and in relation to the discussion in chapter two) under the group of transformations

$$Y \rightarrow \eta_0 Y \tag{4.2}$$

where η_0 is a positive scalar. The vector

$$\omega = Y / \left(Y'Y \right)^{1/2} \tag{4.3}$$

is maximal invariant under this group of transformations. The joint density of ω would be

$$f(\omega) = \frac{1}{2}\Gamma\left(\frac{n}{2}\right)\pi^{-n/2}\left|\Delta(\rho)\right| - \frac{1}{2}\left(\omega'\Delta^{-1}(\rho)\omega\right)^{-n/2} \tag{4.4}$$

with respect to the uniform measure on

$$\left\{ \omega : \omega \in R^n, \omega'\omega = 1 \right\}.$$

Hence the joint density function of ω, under H_a is (4.4), while under H_0 it reduces to

$$f_0(\omega) = \frac{1}{2}\Gamma\left(\frac{n}{2}\right)\pi^{-n/2}.$$

Let us consider the simpler problem of testing H_0 against the simple alternative hypothesis $H_a : \rho = \rho_1 > 0$, in (4.1), where ρ_1 is a known and fixed point. The Neyman-Pearson Lemma yields the critical region of the form

$$\left|\Delta(\rho_1)\right|^{-1/2}\left(\omega'\Delta^{-1}(\rho_1)\omega\right)^{-n/2} \geq c$$

or equivalently,

$$s(\rho_1) = \frac{Y'\Delta^{-1}(\rho_1)Y}{Y'Y} < c_\alpha \tag{4.5}$$

which is the most powerful test, where c_α is an appropriate critical value. For the wider problem of testing H_0 against H_a, the test based on $s(\rho_1)$ in (4.5), is the most powerful invariant at $\rho = \rho_1$ and therefore called is a POI test. The α level critical values can be found by solving the probabilities of the form

$$\Pr\left[s(\rho_1) < c_\alpha \right]$$

$$= \Pr\left[Y'\left(\Delta^{-1}(\rho_1) - c_\alpha I_n \right) Y < 0 \right]$$

$$= \Pr\left[\sum_{i-1}^{n} \lambda_i \xi_i^2 < 0 \right]$$

$$= \Pr\left[\left(\frac{1}{1-\rho_1} - c_\alpha \right) \chi_{m(k-1)}^2 + \left(\frac{1}{1+(k-1)\rho_1} - c_\alpha \right) \chi_m^2 < 0 \right], \qquad (4.6)$$

where $\lambda_i, i=1,\dots,n$, are the eigenvalues of the matrix $\left(\Delta^{-1}(\rho_1) - c_\alpha I_n \right)$, namely $\left(\dfrac{1}{1-\rho_1} - c_\alpha \right)$ and $\left[\dfrac{1}{\{1+(k-1)\rho_1\}} - c_\alpha \right]$ with multiplicities of $m(k-1)$ and m, respectively and $\xi_i \sim IN(0, 1)$.

Alternatively, one can note that the critical values of the $s(\rho_1)$ test can also be obtained by the standard F distribution. Notice that equation (4.6) can be expressed as

$$\Pr\left[s(\rho_1) < c_\alpha \right]$$

$$= \Pr\left[(a - c_\alpha) \chi_{m(k-1)}^2 + (b - c_\alpha) \chi_m^2 < 0 \right]$$

$$= \Pr\left[\left(a - c_\alpha\right) \chi^2_{m(k-1)} < \left(c_\alpha - b\right) \chi^2_m \right]$$

$$= \Pr\left[\frac{\chi^2_{m(k-1)/m(k-1)}}{\chi^2_m / m} < \frac{\left(c_\alpha - b\right)}{\left(a - c_\alpha\right)\left(k - 1\right)} \right]$$

$$= \Pr\left[F_{(m(k-1),m)} < F_\alpha \right],$$

where

$$F_\alpha = \frac{\left(c_\alpha - b\right)}{\left(a - c_\alpha\right)\left(k - 1\right)},$$

$$a = \frac{1}{1 - \rho_1} \text{ and } b = \frac{1}{\left\{1 + \left(k - 1\right)\rho_1\right\}},$$

and $F_{(m(k-1),m)}$ is the standard F distribution with $m(k-1)$ and m degrees of freedom. If F_α is the $100(1-\alpha)$ percentile of the F distribution (obtainable from the standard tables), then the critical values c_α can be calculated by,

$$c_\alpha = \frac{\left\{b + a\left(k - 1\right)F_\alpha\right\}}{\left\{1 + \left(k - 1\right)F_\alpha\right\}}.$$

Thus for any critical value c_α the power of the $s(\rho_1)$ test is obtained by solving probabilities of the form

$$\Pr\left[s\left(\rho_1\right) < c_\alpha \middle| Y \sim N\left(0, \sigma^2 \Delta\left(\rho\right)\right) \right]$$

$$= \Pr\left[Y'\left(\Delta^{-1}\left(\rho_1\right) - c_\alpha I_n\right)Y < 0 \middle| H_a \right]$$

$$= \Pr\left[z'T'\left(\Delta^{-1}\left(\rho_1 \right) - c_\alpha I_n \right) Tz < 0 \right]$$

$$= \Pr\left[\sum_{i=1}^{n} \lambda_i \xi_i^2 < 0 \right], \tag{4.7}$$

where λ_i, $i = 1, 2, ..., n$, are the eigenvalues of

$$T'\left(\Delta^{-1}\left(\rho_1 \right) - c_\alpha I_n \right) T \text{ and } \xi = \left(\xi_1, ..., \xi_n \right)' \sim IN\left(0, I_n \right).$$

The second last equality of (4.7) follows from the Cholesky decomposition of $\Delta(\rho) = TT'$, where the matrix T is given by (3.9) in chapter three. Both (4.6) and (4.7) can be evaluated by standard numerical algorithms, which were mentioned earlier.

As in chapter three, the results of appendix 3A can be used to obtain the eigenvalues of the matrix $T'\left(\Delta^{-1}\left(\rho_1 \right) - c_\alpha I_n \right) T$, in (4.7), analytically. Hence, note that the eigenvalues of $T'\left(\Delta^{-1}\left(\rho_1 \right) - c_\alpha I_n \right) T$ are also the eigenvalues of $\left(\Delta^{-1}\left(\rho_1 \right) - c_\alpha I_n \right) \Delta\left(\rho \right)$, namely

$$= \left(\frac{1}{1-\rho_1} - c_\alpha \right)\left(1-\rho \right) I_n + \left[\left(\frac{1}{1-\rho_1} - c_\alpha \right)\rho - \frac{\rho_1\left(1-\rho \right) + \rho_1\rho k}{\left\{ 1 + \rho_1\left(k-1 \right) \right\}\left(1-\rho_1 \right)} \right] D$$

$$= a * I_n + b * D$$

These eigenvalues are:

$$a* = \left(1-\rho \right)\left(\frac{1}{1-\rho_1} - c_\alpha \right) \tag{4.8}$$

and

$$a*+b*k = \left[1+\rho(k-1)\right]\left[\frac{1}{1+\rho_1(k-1)}-c_\alpha\right]$$ (4.9)

with multiplicities of $m(k-1)$ and m respectively.

Thus one can write (4.7) under (4.8) and (4.9) as

$$\Pr\left[a*\chi^2_{m(k-1)}+\left(a*+b*k\right)\chi^2_m <0\right]$$

$$= \Pr\left[a*\chi^2_{m(k-1)} <-\left(a*+b*k\right)\chi^2_m\right]$$

$$= \Pr\left[\frac{\chi^2_{m(k-1)/m(k-1)}}{\chi^2_{m/m}} < \frac{-\left(a*+b*k\right)}{a*\left(k-1\right)}\right]$$

$$= \Pr\left[F_{\left(m(k-1),m\right)} < F_p\right]$$

where p denotes the power and F_p is obtained from

$$F_p = \frac{-\left(a*+b*k\right)}{a*\left(k-1\right)},$$

where $a*$ and $(a* + b*k)$ are given by (4.8) and (4.9) respectively.of the features of the $s(\rho_1)$ test, as noted earlier, is that its power and critical values are obtainable from the standard F distribution, with the $s(\rho_1)$ test proved below to be UMPI test proved in theorem (4.1).

Theorem 4.1.

The $s(\rho_1)$ test is a UMPI test.

Proof:

Consider equation (4.5), i.e.,

$$s(\rho_1) = \frac{Y'\Delta^{-1}(\rho_1)Y}{Y'Y} < c_\alpha$$

where $\Delta^{-1}(\rho_1)$ in (3.6) can be written as:

$$\Delta^{-1}(\rho_1) = aI_n + bD,$$

with

$$a = \frac{1}{1-\rho_1},$$

$$b = \frac{-\rho_1}{(1-\rho_1)\{1+(k-1)\rho_1\}}, \text{ then}$$

$$s(\rho_1) = \frac{Y'(aI_n + bD)Y}{Y'Y} < c_\alpha$$

$$s(\rho_1) = a + b\frac{Y'DY}{Y'Y} < c_a$$

or equivalently

$$\frac{Y'DY}{Y'Y} > d_\alpha,$$

where

$$d_\alpha = \frac{c_\alpha - a}{b}.$$

Note that in the above inequality b is negative when it is multiplied by $1/b$ the inequality reverses. Expressed in this form, the critical value d_α is obtained by reference to the null distribution of $\dfrac{Y'DY}{Y'Y}$. Hence the critical region of the $s(\rho_1)$ test can be written independent of ρ_1, and therefore the test is UMPI.

4.3. A MORE COMPLICATED MODEL AND THE LMMPI TEST

Consider an $n \times 1$ dimensional random vector Y such that

$$Y \sim N\left(0, \sigma^2 \Delta(\rho)\right),$$

where $\Delta(\rho)$ is given as (3.15), with the addition of σ^2 as an unknown scalar and $\rho = (\rho_1, \ldots, \rho_m)'$. Here the problem of interest is to test the null hypothesis $H_0 : \rho_1 = \ldots = \rho_m = 0$, against $H_a : \rho_i \geq 0$, for $i=1,2,\ldots,m$. Note that this testing problem is invariant under the group of transformations (4.2), with the maximal invariant (4.3), and the density of the maximal invariant given by (4.4). Let

$$A_i = -\left.\frac{\partial \Delta(\rho)}{\partial \rho_i}\right|_{\rho=0} = \left.\frac{\partial \Delta^{-1}(\rho)}{\partial \rho_i}\right|_{\rho=0},$$

where $\Delta^{-1}(\rho)$ is given by (3.18), and the elements of the matrix A_i are given by (3.21).

Using equation (3) of King and Wu's (1990) theorem and the density (4.4), the LMMPI test for this model is based on the critical region of the form

$$\sum_{i=1}^{m} -\frac{1}{2}\left.\frac{\partial \ell n |\Delta(\rho)|}{\partial \rho_i}\right|_{\rho=0} - \sum_{i=1}^{m} \frac{n}{2}\left.\frac{\partial \ell n\left(\omega'\Delta^{-1}(\rho)\omega\right)}{\partial \rho_i}\right|_{\rho=0} > c.$$

The first summation on the left is simply a scalar constant. Evaluating the second summation gives the critical region of the form

$$-\frac{1}{2}c_1 - \frac{n}{2}\sum_{i=1}^{m}\omega' A_i \omega > c$$

or in a simplified form,

$$s = \sum_{i=1}^{m}s_i = \sum_{i=1}^{m}\frac{Y'A_iY}{Y'Y} = \frac{Y'AY}{Y'Y}$$

or equivalently,

$$s = \frac{Y'DY}{Y'Y} > (1-c_2), \qquad (4.10)$$

where $A = (I_n - D)$ and c_2 is a suitable chosen constant. Note that (4.10) is also the UMPI test of testing H_0 against H_a, and it can further be simplified as,

$$\frac{\sum_{i=1}^{m}\left(\sum_{j=1}^{k}y_{ij}\right)^2}{\sum_{i=1}^{m}\sum_{j=1}^{k}y_{ij}^2} > (1-c_2)$$

with its critical values and the power also obtained via standard F distribution, in a similar way to that of the $s(\rho_1)$ test in the previous section.

4.4. THREE-STAGE SMC MODEL AND THE TESTS

This section extends the two-stage SMC model of the previous section to the situation in which observations are obtained from a three-stage cluster design, i.e.,

$$Y \sim N\left(0, \sigma^2 \Omega\left(\rho_1, \rho_2\right)\right), \tag{4.11}$$

where the matrix form Ω (ρ_1, ρ_2) in (4.11) is given by (3.26), with the addition of the unknown constant σ^2.

Locally most mean powerful invariant test.

In this subsection the null hypothesis under consideration is of,

$$H_0 : \rho_1 = \rho_2 = 0,$$

which is tested against an alternative hypothesis

$$H_a : \rho_1 \geq 0, \rho_2 \geq 0, \text{ (excluding } H_0).$$

The above testing problem is also invariant under the group of transformations (4.2), with the vector ω, given by (4.3), is a maximal invariant vector, whose joint density is (4.4), with the modification that the covariance matrix $\Delta(\rho)$ is replaced by $\Omega(\rho_1, \rho_2)$. This the LMMPI test for the three-stage model is based on the critical region of the form

$$d = \frac{Y'AY}{Y'Y} < c_\alpha, \tag{4.12}$$

where c is a suitably chosen constant. Note that in (4.12) above, the matrix A can be expressed in a simplified form as

$$A = A_1 + A_2 = (I_n - D_1) + (I_n - D_2) = 2I_n - (D_1 + D_2),$$

where,

$$A_1 = \left(I_n - D_1\right),$$
$$A_2 = \left(I_n - D_2\right)$$

And where matrices D_1 and D_2 in (4.12) above are defined earlier in (3.13) under the restriction that the cluster and subcluster sizes are balanced.

More conveniently, d in (4.12) can also be expressed as

$$d \;=\; \frac{Y' A_1 Y}{Y'Y} + \frac{Y' A_2 Y}{Y'Y}$$

$$= \frac{Y'\left(I_n - D_1\right)Y}{Y'Y} + \frac{Y'\left(I_n - D_2\right)Y}{Y'Y}$$

$$= 2 - \left[\frac{Y'D_1 Y}{Y'Y} + \frac{Y'D_2 Y}{Y'Y}\right]$$

$$= 2 - \left[\sum_{i=1}^{m}\left(\sum_{j=1}^{s}\sum_{k=1}^{T} y_{ijk}\right)^2 + \sum_{i=1}^{m}\sum_{j=1}^{s}\left(\sum_{k=1}^{T} y_{ijk}\right)^2\right] \Big/ \sum_{i=1}^{m}\sum_{j=1}^{s}\sum_{k=1}^{T} y_{ijk}^2 . \quad (4.13)$$

Note that d in (4.12) and/or in (4.13) is a ratio of quadratic forms in normal variables, and therefore its critical values, c_α, may be obtained by evaluating probabilities of the form

$$\Pr\left[d < c_\alpha\right] = \Pr\left[Y'\left(A - c_\alpha I_n\right)Y < 0\right]$$

$$= \Pr\left[Y'\left\{\left(2 - c_\alpha\right)I_n - \left(D_1 + D_2\right)\right\}Y < 0\right]$$

$$= \Pr\left[\sum_{i=1}^{n} \lambda_i \xi_i^2 < 0\right]$$

$$= \Pr\left[\left(2 - c_\alpha\right)\chi_{ms(T-1)}^2 + \left(2 - c_\alpha - T\right)\chi_{m(s-1)}^2\right.$$

$$\left. + \left(2 - c_\alpha - T(s+1)\right)\chi_m^2 < 0\right], \qquad (4.14)$$

where $\xi_i \sim IN(0, 1)$ and λ_i, $i = 1, 2,...,n$, are eigenvalues of the matrix $\{(2 - c_\alpha)I_n - (D_1 + D_2)\}$. The last equality of (4.14) is obtained by noting that the eigenvalues of the latter matrix are $(2 - c_\alpha)$ with multiplicity of $ms(T - 1)$, $(2 - c_\alpha - T)$, with multiplicity of $m(s-1)$ and $\{2 - c_\alpha - T(s + 1)\}$ with multiplicity of m, respectively. Also (4.14) can be evaluated by standard numerical algorithms, which has been mentioned previously.

Point Optimal Invariant Test

This subsection reconsiders the testing problem of the previous subsection, i.e. $H_0 : \rho_1 = \rho_2 = 0$ against the simple alternative $H_a : \rho_1 = \rho_{11} > 0, \rho_2 = \rho_{22} > 0$.

Based on the maximal invariant (4.3), King's (1987b) equation (18) gives the POI test of H_0 against H_a. It rejects the null hypothesis for small values of

$$s(\rho_{11}, \rho_{22}) = \frac{Y'\Omega^{-1}(\rho_{11}, \rho_{22})Y}{Y'Y},$$

where $\Omega^{-1}(\rho_{11}, \rho_{22})$ is given by (3.33).

Note that $s(\rho_{11}, \rho_{22})$ is a ratio of quadratic forms in normal variables, and therefore its critical values may be obtained by evaluating the probabilities of the form

$$\Pr\left[s(\rho_1, \rho_2) < c_\alpha\right] = \Pr\left[Y'\left(\Omega^{-1}(\rho_{11}, \rho_{22}) - c_\alpha I_n\right)Y < 0\right]$$

$$= \Pr\left[\sum_{i=1}^{n} \lambda_i \xi_i^2 < 0\right]$$

$$= \Pr\left[\left\{(1 - \rho_{11} - \rho_{22})^{-1} - c_\alpha\right\}\chi_{ms(T-1)}^2\right]$$

$$+\left\{\left(1-\rho_{11}-\rho_{22}\left(1-T\right)\right)^{-1}-c_{\alpha}\right\}\chi^{2}_{m(s-1)}$$

$$+\left\{\left(1+\rho_{11}\left(sT-1\right)+\rho_{22}\left(T-1\right)\right)^{-1}-c_{\alpha}\right\}\chi^{2}_{m}<0\right],\qquad(4.15)$$

where $\xi_{i}\sim IN\left(0,1\right)$ and $\lambda_{i}, i=1,\ldots,n,$ are the eigenvalues of the matrix $\left(\Omega^{-1}\left(\rho_{11},\rho_{22}\right)-c_{\alpha}I_{n}\right),$

namely,

$$\left\{\left(1-\rho_{11}-\rho_{22}\right)^{-1}-c_{\alpha}\right\},$$

$$\left\{\left(1-\rho_{11}-\rho_{22}\left(1-T\right)\right)^{-1}-c_{\alpha}\right\}$$

and

$$\left\{\left(1+\rho_{11}\left(sT-1\right)+\rho_{22}\left(T-1\right)\right)^{-1}-c_{\alpha}\right\},$$

with multiplicities of $ms(T-1)$, $m(s-1)$ and m respectively.

Given a critical value, c_{α}, for a desired level of significance, α, the power of the critical region $s(\rho_{11}, \rho_{22}) < c_{\alpha}$ is obtained by evaluating the probabilities of the form

$$\Pr\left[s\left(\rho_{11},\rho_{22}\right)<c_{\alpha}\middle|Y\sim N\left(0,\sigma^{2}\Omega\left(\rho_{1},\rho_{2}\right)\right)\right]$$

$$=\Pr\left[Y'\left(\Omega^{-1}\left(\rho_{11},\rho_{22}\right)-c_{\alpha}I_{n}\right)Y<0\middle|Y\sim N\left(0,\sigma^{2}\Omega\left(\rho_{1},\rho_{2}\right)\right)\right]$$

$$=\Pr\left[z'\left(\Omega^{1/2}\right)'\left(\Omega^{-1}\left(\rho_{11},\rho_{22}\right)-c_{\alpha}I_{n}\right)\Omega^{1/2}z<0\right]$$

$$= \Pr\left[\sum_{i=1}^{n} \lambda_i \xi_i^2 < 0\right], \tag{4.16}$$

where $\xi_i \sim IN(0,1), \lambda_i, i = 1,...,n,$ are the eigenvalues of $\left[\left(\Omega^{1/2}\right)'\left(\Omega^{-1}\left(\rho_{11},\rho_{22}\right) - c_\alpha I_n\right)\Omega^{1/2}\right]$. The second last equality of (4.16) is obtained by noting that $z = \Omega^{-1/2} Y \sim N(0, I_n)$, where $\Omega = \Omega(\rho_1, \rho_2)$.

This follows from lemma 2 of appendix 3A, that the eigenvalues of

$$\left[\left(\Omega^{1/2}\right)'\left(\Omega^{-1}\left(\rho_{11},\rho_{22}\right) - c_\alpha I_n\right)\Omega^{1/2}\right]$$

are also the eigenvalues of

$$\left(\Omega^{-1}\left(\rho_{11},\rho_{22}\right) - c_\alpha I_n\right)\Omega(\rho_1,\rho_2),$$

or equivalently, the eigenvalues of

$$\left\{\Omega^{-1}\left(\rho_{11},\rho_{22}\right)\Omega(\rho_1,\rho_2) - c_\alpha \Omega(\rho_1,\rho_2)\right\}.$$

The eigenvalues of the latter matrices are obtained by using lemma 4 of appendix 3A. These eigenvalues are

$$a^* = (1 - \rho_1 - \rho_2)\left[\frac{1}{1 - \rho_{11} - \rho_{22}} - c_\alpha\right], \tag{4.17}$$

$$b^* = \left\{1 - \rho_1 - \rho_2(1 - T)\right\}\left[\frac{1}{\left\{1 - \rho_{11} - \rho_{22}(1 - T)\right\}} - c_\alpha\right], \tag{4.18}$$

and

$$c^* = \left\{1 - \rho_1\left(1-sT\right) - \rho_2\left(1-T\right)\right\}\left[\frac{1}{\left\{1 - \rho_{11}\left(1-sT\right) - \rho_{22}\left(1-T\right)\right\}} - c_\alpha\right],$$

(4.19)

with multiplicities of $ms(T-1)$, $m(s-1)$ and m, respectively, thus, giving these eigenvalues, the power function (4.16), which can now be written as

$$\Pr\left[a^* \chi^2_{ms(T-1)} + b^* \chi^2_{m(s-1)} + c^* \chi^2_m < 0\right],$$

(4.20)

where a^*, b^* and c^* are defined in (4.17), (4.18) and (4.19), respectively. In the special cases in which power is being evaluated at

$$\left(\rho_1, \rho_2\right)' = \left(\rho_{11}, \rho_{22}\right)',$$
(4.17), (4.18) and (4.19) become

$$1 - c_\alpha\left(1 - \rho_{11} - \rho_{22}\right),$$

$$1 - c_\alpha\left(1 - \rho_{11} - \rho_{22}\left(1-T\right)\right),$$

and

$$1 - c_\alpha\left\{1 - \rho_{11}\left(1-sT\right) - \rho_{22}\left(1-T\right)\right\},$$

respectively. Thus (4.20) can be written as

$$\Pr\left[\left\{1 - c_\alpha\left(1 - \rho_{11} - \rho_{22}\right)\right\}\chi^2_{ms(T-1)} + \left\{1 - c_\alpha\left(1 - \rho_{11} - \rho_{22}\left(1-T\right)\right)\right\}\right.$$

$$\left.\chi^2_{m(s-1)} + \left[1 - c_\alpha\left\{1 - \rho_{11}\left(1-sT\right) - \rho_{22}\left(1-T\right)\right\}\right]\chi^2_m < 0\right]$$

$$= \Pr\left[\chi^2_{ms(T-1)} + d\chi^2_{m(s-1)} + f\chi^2_m < 0\right],$$

(4.21)

where

$$d = \frac{\left\{1 - c_\alpha \left(1 - \rho_{11} - \rho_{22}\left(1 - T\right)\right)\right\}}{\left\{1 - c_\alpha \left(1 - \rho_{11} - \rho_{22}\right)\right\}}$$

and

$$f = \frac{\left[1 - c_\alpha \left\{1 - \rho_{11}\left(1 - sT\right) - \rho_{22}\left(1 - T\right)\right\}\right]}{\left\{1 - c_\alpha \left(1 - \rho_{11} - \rho_{22}\right)\right\}}.$$

Note that both (4.16) and (4.21) can be evaluated by using standard numerical techniques mentioned in previous sections. In using the same arguments as previously stated, it can be easily seen that the $s(\rho_{11}, \rho_{22})$ test is a POI for the three-stage SMC distribution model.

4.5. POI TEST FOR TESTING SUBCLUSTER EFFECTS

In this section the similar testing problem to that of testing (3.41) against (3.42) will be considered under the restriction that Y follows SMC distributions model, which is given by (4.11). This testing problem is invariant under the group of transformations of the form (4.2), with maximal invariant vector, ω, is given by (4.3). The density of ω, under H_a is given by:

$$f_a\left(\omega, \rho_1, \rho_2\right) = \frac{1}{2}\Gamma\left(\frac{n}{2}\right)\pi^{-n/2}\left|\Omega_1\right|^{-1/2}\left(\omega'\Omega_1^{-1}\omega\right)^{-n/2},$$

while under H_0 it becomes:

$$f_0\left(\omega, \rho_1, 0\right) = \frac{1}{2}\Gamma\left(\frac{n}{2}\right)\pi^{-n/2}\left|\Omega_0\right|^{-1/2}\left(\omega'\Omega_0^{-1}\omega\right)^{-n/2}\int$$

where Ω_0 and Ω_1 are defined earlier in chapter three. Thus, our testing problem (3.41) against (3.42) simplifies to one of testing

$$H_0 : \omega \text{ has the density } f_0\left(\omega, \rho_1, 0\right), \ 0 \leq \rho_1 \leq 0.5,$$

against

$$H_a : \omega \text{ has the density } f_a\left(\omega, \rho_1, \rho_2\right), \ 0 \leq \rho_1 \leq 0.5, 0 \leq \rho_2 \leq 0.5.$$

To begin with, the simpler problem of testing

$$H_0' : \left(\rho_1, \rho_2\right)' = \left(\rho_{10}, 0\right)'$$

against the simple alternative hypothesis

$$H_a' : \left(\rho_1, \rho_2\right)' = \left(\rho_{11}, \rho_{21}\right)',$$

where the range of ρ_{10}, ρ_{11}, ρ_{21} are the same as defined earlier in chapter three, and dealing with the similar problem of SSMC model. As usual, in the previous subsections, the Neyman-Pearson lemma yields the critical region of the form

$$s\left(\rho_{10}, \rho_{11}, \rho_{21}\right) = \frac{Y'\Omega_{11}^{-1}Y}{Y'\Omega_{10}^{-1}Y} < c_\alpha \tag{4.22}$$

where c_α in (4.22) is an appropriate critical value. In order to compute c_α, one needs to evaluate the probability of the form

$$\Pr\left[s\left(\rho_{10}, \rho_{11}, \rho_{21}\right) < c_\alpha \middle| Y \sim N\left(0, \Omega\left(\rho_{10}, 0\right)\right) \right]$$

$$= \Pr\left[Y'\left(\Omega_{11}^{-1} - c_\alpha \Omega_{10}^{-1}\right)Y < 0 \mid Y \sim N\left(0, \Omega_0\right) \right]$$

$$= \Pr\left[z'\left(\Omega_0^{1/2}\right)'\left(\Omega_{11}^{-1} - c_\alpha\Omega_{10}^{-1}\right)\Omega_0^{1/2}z < 0 \mid z \sim N\left(0, I_n\right)\right]$$

$$= \Pr\left[\sum_{i=1}^{n} \lambda_i \xi_i^2 < 0\right] \qquad\qquad (4.23)$$

where $\xi_i \sim IN(0,1), \lambda_i, i = 1,..., n,$ are the eigenvalues of

$$\left(\Omega_0^{1/2}\right)'\left(\Omega_{11}^{-1} - c_\alpha\Omega_{10}^{-1}\right)\Omega_0^{1/2},$$

or equivalently, the eigenvalues of

$$\left(\Omega_{11}^{-1} - c_\alpha\Omega_{10}^{-1}\right)\Omega_0 = \Omega_{11}^{-1}\Omega_0 - c_\alpha\Omega_{10}^{-1}\Omega_0,$$

in which $\Omega_{11} = \Omega\left(\rho_{11}, \rho_{22}\right), \Omega_{10} = \Omega\left(\rho_{10}, 0\right)$ and $\Omega_0 = \Omega\left(\rho_1, 0\right)$.

Using lemmas 2 and 4 of appendix 3A, the eigenvalues of the latter matrix can be obtained as

$$\bar{a} = \left(1 - \rho_1\right)\left\{\frac{1}{1 - \rho_{11} - \rho_{22}} - \frac{c_\alpha}{1 - \rho_{10}}\right\}, \qquad\qquad (4.24)$$

$$\bar{b} = \frac{\left(1 - \rho_1\right)}{\left(1 - \rho_{11} - \rho_{22}\right)}\left\{\frac{1 - \rho_{10} - \rho_{22}}{1 - \rho_{11} - \rho_{22} + T\rho_{22}} - \frac{c_\alpha}{1 - \rho_{10}}\right\}, \qquad\qquad (4.25)$$

and

$$\bar{c} = \left(1 - \rho_1 + sT\rho_1\right)\left\{\frac{1}{\left(1 - \rho_{11} - \rho_{22} + T\rho_{22} + sT\rho_{11}\right)}\right.$$

$$\left. -c_\alpha\left(\frac{1}{\left(1 - \rho_{10}\right)\left(1 - \rho_1 + sT\rho_1\right)} - \frac{\rho_{10}}{\left(1 - \rho_{10}\right)\left(1 - \rho_{10} + sT\rho_{10}\right)}\right)\right\}, \qquad\qquad (4.26)$$

with multiplicities $ms(T-1)$, $m(s-1)$ and m respectively. Therefore (4.23) can be written as

$$\Pr\left[\bar{a}\,\chi^2_{ms(T-1)} + \bar{b}\,\chi^2_{m(s-1)} + \bar{c}\,\chi^2_m\right],$$

where $\bar{a}, \bar{b}$ and $\bar{c}$ are given by (4.24), (4.25) and (4.26). Thus a test based on $s\left(\rho_{10},\rho_{11},\rho_{21}\right)$ in (4.22) is most powerful in the neighbourhood of $\left(\rho_1,\rho_2\right)' = \left(\rho_{11},\rho_{21}\right)'$, and therefore is a POI test of a simple null H_0' against a simple alternative H_a'. For the wider problem of testing (3.41) against (3.42), it may not necessarily be POI, because for such a testing problem one needs to solve the probability of the form (3.50), based on the $s\left(\rho_{10},\rho_{11},\rho_{21}\right)$ test for at least one ρ_1 value. If one could succeed to find such a value of $\rho_1 = \rho_{10}$, say, then (4.24), (4.25) and (4.26) become

$$\bar{a}' = \left\{\frac{1-\rho_{10}}{1-\rho_{11}-\rho_{22}} - c_\alpha\right\},$$

$$\bar{b}' = \left\{\frac{\left(1-\rho_{10}\right)}{\left(1-\rho_{11}-\rho_{22}+T\rho_{22}\right)} - \frac{c_\alpha}{\left(1-\rho_{11}-\rho_{22}\right)}\right\},$$

and

$$\bar{c}' = \left\{\frac{\left(1-\rho_{10}+sT\rho_{10}\right)}{\left(1-\rho_{11}-\rho_{22}+T\rho_{22}+sT\rho_{11}\right)} - c_\alpha\right\}, \text{ respectively.}$$

Thus the critical values can be obtained by solving the left hand side of

$$\Pr\left[\bar{a}'\,\chi^2_{ms(T-1)} + \bar{b}'\,\chi^2_{m(s-1)} + \bar{c}'\,\chi^2_m < 0\right] = \alpha,$$

where α is the desired significance level. The exact critical values and the power calculations for the similar testing problem will be considered in chapter six, while dealing with the 3SLR model in which disturbances follow these distributions.

The LMMPI and the POI tests can easily be generalized to a multi-stage cluster design for testing cluster, sub-cluster and/or sub-sub-cluster equicorrelation coefficients up to multivariate levels, whether these correlation coefficients are equal to zero, or they have some positive values, within clusters, sub-clusters and/or sub-sub-clusters, respectively. Discussion on some of these more complicated models is considered in chapter six of this book.

4.6. CONCLUDING REMARKS

In this chapter the POI and LMMPI tests for the case of two- and three-stage SMC distribution model were discussed. These tests have been developed in the presence of the nuisance parameter, $\sigma 2$, which has been eliminated by using an invariance technique. The most attractive feature of the POI and LMMPI tests for the case of the two-stage SMC distribution model is that their power and the critical values are easily obtainable from the standard F distribution. An additional feature of the POI test is that it is UMPI.

The results of LMMPI and POI tests for the case of three-stage SMC model for testing whether $\rho_1 = \rho_2 = 0$ against the alternative is, that some positive values were derived. The POI test for testing $\rho2$ in the presence of $\rho_1 > 0$ was also developed, with a brief summary of the tests in the previous and this chapter, and summarized in tables 4.1 and 4.2 of appendix 4A. There are some other topics worth exploring when using a model which follows a SSMC or SMC distribution. One of the questions of testing cluster and subcluster effects are developed in the subsequent chapters of the rest of this book in the context of the multi-stage linear regression model, when disturbances follow these distributions.

APPENDIX 4A. SUMMARY OF THE TESTS OF THE SSMC AND SMC MODELS

Table 4.1. Some optimal tests for two-stage SSMC and SMC distributions

Name of tests	Null hypothesis	Alternative hypothesis	Form of covariance matrix	Test statistics
SSMC				
1 LB	$\rho = 0$	$\rho > 0$	$\Delta(\rho_0)$	$s = Y'(I_n - D)Y < c_\alpha$
2 BO	$\rho = 0$	$\rho > 0$	$\Delta(\rho_1)$	$r(\rho_1) = Y'(\Delta^{-1}(\rho_1) - I_n)Y$
3 LMMP	$\rho_i = 0$ $i = 1,...,m$	$\rho_i \geq 0$ $i = 1,...,m$	$\Delta(\rho_i)$	$d = Y'(I_n - D)Y < c_2$
SMC				
4 POI	$\rho = 0$	$\rho > 0$	$\sigma^2 \Delta(\rho_1)$	$s(\rho_1) = \dfrac{Y'\Delta^{-1}(\rho_1)Y}{Y'Y} < c_\alpha$
6 LMMPI	$\rho_i = 0$ $i = 1,...,m$	$\rho_i \geq 0$ $i = 1,...,m$	$\sigma^2 \Delta(\rho_i)$	$s = \dfrac{Y'(I_n - D)Y}{Y'Y} < c_2$

Table 4.2. Some optimal tests for three-stage SSMC and SMC distributions

Name of tests	Null hypothesis	Alternative hypothesis	Form of covariance matrix	Test statistics
SSMC				
1 LMMP	$\rho_1 = \rho_2 = 0$	$\rho_1 \geq 0, \rho_2 \geq 0$	$\Omega(\rho_1, \rho_2)$	$d^* = Y'\{2I_n - (D_1 + D_2)\}Y < c_{3^-}$
2 PO	$\rho_1 = \rho_2 = 0$	$\rho_1 = \rho_{11} > 0$ $\rho_2 = \rho_{22} > 0$	$\Omega(\rho_{11}, \rho_{22})$	$r(\rho_{11}, \rho_{22}) = Y'(\Omega^{-1}(\rho_{11}, \rho_{22}) - 1$
3 PO	$\rho_2 = 0, \rho_1 > 0$	$\rho_1 = \rho_{11} > 0$	under $H_0 : \Omega_{10} = \Omega(\rho_{10}, 0)$	
(for sub-cluster effects)		$\rho_2 = \rho_{21} > 0$	under $H_a : \Omega_{11} = \Omega(\rho_{11}, \rho_2$	$r(\rho_{10}, \rho_{11}, \rho_{21}) = Y'(\Omega_{11}^{-1} - \Omega_{10}^{-1})Y$
SMC				
4 LMMPI	$\rho_1 = \rho_2 = 0$	$\rho_1 \geq 0, \rho_2 \geq 0$	$\sigma^2 \Omega(\rho_1, \rho_2)$	$d = 2 - \dfrac{Y'(D_1 + D_2)Y}{Y'Y} < c_\alpha$
5 POI	$\rho_1 = \rho_2 = 0$	$\rho_1 = \rho_{11} > 0$ $\rho_2 = \rho_{22} > 0$	$\sigma^2 \Omega(\rho_{11}, \rho_{22})$	$s(\rho_{11}, \rho_{22}) = \dfrac{Y'\Omega(\rho_{11}, \rho_{22})Y}{Y'Y} < c_{\cdot}$

Table 4.2. (Continued)

Name of tests	Null hypothesis	Alternative hypothesis	Form of covariance matrix	Test statistics
6 POI (for sub-cluster effects)	$\rho_2 = 0, \rho_1 > 0$	$\rho_1 = \rho_{11} > 0$ $\rho_2 = \rho_{22} > 0$	$\sigma^2 \Omega_{10}$ $\sigma^2 \Omega_{11}$	$s(\rho_{10}, \rho_{11}, \rho_{21}) = \dfrac{Y'\Omega_{11}Y}{Y'\Omega_{10}Y} < c_\alpha$

5. TESTING FOR CLUSTER EFFECTS IN 2SLR MODELS

5.1. INTRODUCTION[1]

It is common practice for survey data to be collected in clusters and the standard linear regression model to be fitted to such data, with the disturbances assumed to be mutually independent. This practice entails a number of pitfalls, which have been discussed in the first two chapters of this book. In this chapter, hypothesis testing problems associated with the linear regression model arising from two-stage cluster survey data are considered. To recall King and Evans (1986) considered a regression model based on a two-stage clustered sampling design, and suggested the use of a one-sided Lagrange multiplier (LM1) test. They showed that the test is LBI for testing for a non-zero equicorrelation coefficient, ρ, within the clusters.

The main purpose of this chapter is to construct a POI test and to investigate whether the power of this test is better than that of the LBI test for small and moderate sample sizes. It has been found that for testing $H_0 : \rho = 0$, against $H_a : \rho > 0$, both the LBI and POI tests are approximately UMPI for some selected cluster sizes, and for such cases the critical values of this test can be approximated by critical values from the standard F distribution. The estimated sizes based on these critical values have a tendency to become closer to the nominal size of 0.05, when the sample size and the value of ρ_1 increase.

---

[1] Part of the findings reported in this chapter appeared in Bhatti (2000).

This chapter also derives a locally mean most powerful invariant (LMMPI) test for testing the hypothesis, $\rho = 0$, against the alternative that ρ varies from cluster to cluster, and finds the LMMPI test as equivalent to that of the LM1 test.

The content of this chapter is divided into the following sections. In the subsequent sections, the 2SLR model is introduced and a procedure for constructing a POI test for testing $\rho = 0$, against a positive value of ρ, is outlined. The application of King and Wu's (1990) LMMPI test to more complicated models is also demonstrated, in which it is assumed that the equicorrelation coefficients vary from cluster to cluster and show that it is equivalent to the LM1 test. In the following section, the critical values of the POI test is approximated using the standard F distribution for selected cluster sizes, with the five versions of the POI tests which optimizes power at $\rho = 0.1$, 0.2, 0.3, 0.4 and 0.5, respectively presented. It is found that the $s(0.3)$ test, which optimizes power at $\rho = 0.3$, is the best overall test. This chapter also reports an empirical power comparison of the $s(0.3)$, LM1, LM2, Durbin Watson (DW), and King's (1981) modified DW tests with that of the power envelope (PE), whilst the final section containing some concluding remarks.

5.2. THE 2SLR MODELS AND THE TESTS

Theory

Following King and Evans (1986) and Bhatti (2000), 2SLR model in chapter two, it is assumed that n observations are available from a two-stage sample with m clusters. Let $m(i)$ be the number of observations from the ith cluster so that $n = \sum_{i=1}^{m} m(i)$. It is also assumed that the data is ordered so that the observations within a cluster are adjacent. Therefore, the model in detail is expressed as

$$y_{ij} = \sum_{k=1}^{p} \beta_k x_{ijk} + u_{ij}$$

$$(5.1)$$

where

$$u_{ij} = v_i + v_{ij}, i = 1,...,m, j = 1,2,...,m(i),$$

in which i is the cluster number, j is the observation number in the given cluster, β_k are unknown constant coefficients[2] and x_{ijk} for $k = 1,2,...,p$ are observations on p independent variables, the first of which is a constant. The error terms u_{ij} have two components: v_i, the ith random shock, specific to the ith cluster and the remaining part, and v_{ij}, the individual shocks. Hence it is assumed:

1 v_i and v_{ij} are normal and mutually independent,

2 $E(v_i) = E(v_{ij}) = 0$, for $i = 1,...,m, j = 1,2,...,m(i)$,

3 $E(v_i v_i') = \begin{cases} \sigma_1^2, & i=i' \\ 0, & \text{otherwise,} \end{cases}$

4 $E(v_{ij} v_{i'j'}) = \begin{cases} \sigma_2^2, & i=i', j=j' \\ 0, & \text{otherwise.} \end{cases}$

The above assumptions (1 to 4) imply that

$E(u_{ij}) = 0$, for all i and j;

$$E(u_{ij} u_{st}) = \sigma^2 \delta_{is} \left\{ \rho + (1-\rho)\delta_{jt} \right\}, \tag{5.2}$$

where $\rho = \dfrac{\sigma_1^2}{\sigma^2}$, such that $0 \le \rho \le 1$, $\sigma^2 = \sigma_1^2 + \sigma_2^2$, $\delta_{ii} = 1$ and $\delta_{ij} = 0$, for $i \ne j$. In matrix form, (5.1) can be written as

$$y(i) = X(i)\beta + u(i); \tag{5.3}$$

---

[2] Varying coefficient model is considered by Bhatti (2004), Chapter 23, pp 385-399 in Statistical Data Mining and Knowledge Discovery, edited by H. Bozadogan Chapmann Hall and CRC.

for $i = 1,...,m$, where $y(i)$ is an $m(i)\mathrm{x}1$ vector of y_{ij} values, $X(i)$ is a $m(i)\mathrm{x}p$ matrix whose (j,k)th element is x_{ijk} and $u(i)$ is an $m(i)\mathrm{x}1$ vector of u_{ij} values. If y and u denote the $n\mathrm{x}1$ vectors of stacked $y(i)$ and $u(i)$ vectors, and X is the $n\mathrm{x}p$ stacked matrix of $X(i)$ matrices, then (5.3) can be written as

$$y = X\beta + u \tag{5.4}$$

such that

$$u \sim N\left(0, \sigma^2 \Delta(\rho)\right),$$

where $\Delta(\rho) = \overset{m}{\underset{i=1}{\oplus}} \Delta_i(\rho)$ is cluster diagonal, with submatrices

$$\Delta_i(\rho) = (1-\rho)I_{m(i)} + \rho E_{m(i)} \tag{5.5}$$

in which $I_{m(i)}$ is a $m(i)\mathrm{x}m(i)$ identity matrix and $E_{m(i)}$ is $m(i)\mathrm{x}m(i)$ matrix, with all elements equal to one. Although we can proceed with unequal cluster sizes in this model, for the sake of simplicity in exposition, we will consider the case where there is balanced data, (i.e., equal cluster sizes) and assume that $m(i) = k$ and $n = km$ in the subsequent sections. With this notation, one can note that (5.5) is equivalent to (3.1), and hence $\Delta(\rho)$ in this chapter is the same as in (3.4) of chapter three.

Point Optimal Invariant Test

In this subsection of the chapter we construct the MPI test statistic for testing H_0: $\rho = 0$ against the specific alternative H_a' : $\rho = \rho_1 > 0$, in the context of model (5.1), and under the assumptions (1 to 4). This statistic forms the basis of a point optimal test for the wider testing problem.

Observe that the problem of testing H_0:$\rho = 0$, against H_a:$\rho > 0$ is invariant under the group of transformations

$$y \rightarrow \gamma_0 y + X\gamma, \tag{5.6}$$

where γ_0 is a positive scalar and γ is $p \times 1$ vector. Let $M_x = I_n - X(X'X)^{-1}X'$, $\hat{u}$ $= M_x y = M_x u$ be the OLS residual vector from (5.4) and P be an $(n-p) \times n$ matrix such that $PP' = I_{(n-p)}$ and $P'P = M_x$. The vector

$$\vartheta = \frac{P\hat{u}}{\left(\hat{u}'P'P\hat{u}\right)^{1/2}}$$

is a maximal invariant under the group of transformations given by (5.6). Based on ϑ, we choose $\rho = \rho_1$, a known point at which we want to have optimal power in the alternative parameter space. By using previous chapters arguments, we can obtain the POI test, which is to reject the null hypothesis for small values of

$$s(\rho_1) = \frac{\hat{u}'A\hat{u}}{\hat{u}'\hat{u}},$$

$$= \frac{u'Au}{u'M_x u}, \tag{5.7}$$

where

$$A = \Delta_1^{-1} - \Delta_1^{-1} X \left(X'\Delta_1^{-1}X\right)^{-1} X'\Delta_1^{-1},$$

$$\Delta_1 = \Delta\left(\rho_1\right).$$

The last equality of (5.7) is obtained by observing that $\hat{u}'A\hat{u} = u'M_x A M_x u = u'Au$. As King (1987a, p. 23) notes, $s(\rho_1)$ in (5.7) can also be written as

$$\frac{\sum_{i=1}^{n-p} \lambda_i \xi_i^2}{\sum_{i=1}^{n-p} \xi_i^2} \tag{5.8}$$

where $\lambda_1, ..., \lambda_{n-p}$ are the reciprocals of the non-zero eigenvalues of $\Delta_1 M_x$ or, equivalently the non-zero eigenvalues of A. Thus, for any given value of ρ_1, at the desired level of significance, α, the critical value, c_α can be found by solving

$$\Pr\left[\sum_{i=1}^{n-p}(\lambda_i - c_\alpha)\xi_i^2 < 0\right] = \alpha. \tag{5.9}$$

This can be done by using either standard methods via methods or existing subroutines of Koerts and Abrahamse's (1969) FQUAD subroutine, Davies' (1980) versions of Imhof's (1961) algorithm or Farebrother's (1980) version of Pan's (1968) procedure to evaluate the left-hand-side of (5.9). Moreover one can also use, Shively, Ansley and Kohn (1990) and Palm and Sneek (1984) alternative procedures to compute (5.9), which do not require the computation of the eigenvalues.

Once these critical values have been found, one may wish to calculate the power of the $s(\rho_1)$ test. To do so, one needs to decompose the error covariance matrix $\Delta(\rho)$, or its diagonal component matrices $\Delta_i(\rho)$, in (5.5). This can be done by noting that the Cholesky decomposition of $\Delta(\rho)$ is

$$\Delta(\rho) = T(\rho)T'(\rho)$$

where $T(\rho)$ can be obtained through S, via the recursive scheme given in (3.9). Observe that the power of the critical region $s(\rho_1) < c_\alpha$ in (5.7) can be found by evaluating the probabilities of the form

$$\Pr(s < c_\alpha) = \Pr\left[u'(A - c_\alpha M_x)u < 0\right], \tag{5.10}$$

where

$$u \sim N\left(0, \sigma^2 \Delta(\rho)\right).$$

Let $\xi = T(\rho)^{-1} u \sim N\left(0, \sigma^2 I_n\right)$, so that $u = T(\rho)\xi$. Substituting for u in (5.10), one can obtain the power function of the form

$$\Pr\left[\xi' T'(\rho)\left(A - c_\alpha M_x\right) T(\rho)\xi < 0\right]$$

$$= \Pr\left[\xi' F \xi < 0\right]$$

$$= \Pr\left[\sum_{i=1}^{n} \lambda_i \eta_i^2 < 0\right], \tag{5.11}$$

where the λ_i's are the eigenvalues of $F = T'(\rho)\left(A - c_\alpha M_x\right) T(\rho)$, and the η_i^2's are independent chi-squared random variables each with one degree of freedom. It is worth noting that under H_a the matrix A in (5.7), can be expressed as

$$A = \left(T^{-1}(\rho_1)\right) M * T^{-1}(\rho_1),$$

where

$$M* = I - X*\left(X*' X*\right)^{-1} X*',$$

and

$$X* = T^{-1}(\rho_1)X.$$

The cluster diagonal lower triangular matrix

$T^{-1}(\rho)$ is of the form[3]

$$T^{-1}(\rho) = \begin{bmatrix} Q_k & 0 & \cdots & 0 \\ 0 & Q_k & \cdots & \cdot \\ \cdot & & & \cdot \\ & & \cdot & \cdot \\ \cdot & & & \cdot \\ \cdot & & & \cdot \\ 0 & \cdots & 0 & Q_k \end{bmatrix}$$

where Q_k is a $k \times k$ lower triangular matrix whose elements are obtainable by the formula

$$q_{ii} = \sqrt{\frac{\left[1+(i-2)\rho\right]}{(1-\rho)\left[1+(i-1)\rho\right]}}$$

and

$$q_{ij} = \frac{-\rho}{\sqrt{(1-\rho)\left[1+(i-1)\rho\right]\left[1+(i-2)\rho\right]}}$$

for $i = 1, 2,...,k$; $j = 2, 3,...,k$ and $j > i$. The last equality of (5.12) can be calculated by using any of the methods outlined above for calculating (5.9).

Locally Most Mean Powerful Invariant Test

In previous subsections a POI test was constructed for the equicorrelation coefficient, ρ, which is based on the error covariance matrix (5.5) of model

---

[3] Note that T-1(ρ) obtained by Cholesky decomposition of the covariance matrix $\Delta^{-1}(\rho) = (T^{-1}(\rho))'T^{-1}(\rho)$ is not unique.

(5.4), and involves only one parameter, ρ, in all the clusters. In this subsection a more general situation is considered which is based on the discussion in chapter three, where it is assumed that the equicorrelation coefficient varies from cluster to cluster. Therefore, in such cases the subcluster of the error covariance matrix (5.5) can be expressed in the form of (3.5), i.e.

$$\Delta_i\left(\rho_i\right) = \left(1 - \rho_i\right)I_k + \rho_i E_k,$$

for $i = 1,2,...,m$.

Our problem of interest is to test

$$\overline{H}_0 : \rho_1 = \rho_2 = ... = \rho_m = 0 \tag{5.12}$$

against

$$\overline{H}_a : \rho_1 \geq 0, \rho_2 \geq 0,..., \rho_m \geq 0, \text{ (excluding } H_0). \tag{5.13}$$

Note that this testing problem is invariant under the group of transformations given by (5.6).

As noted in earlier chapters, the LMMPI test for $\overline{H}_0$ against $\overline{H}_a$ is to reject $\overline{H}_0$ for small values of

$$d' = \frac{\hat{u}'A*\hat{u}}{\hat{u}'\hat{u}}, \tag{5.14}$$

where

$$A* = \sum_{j=1}^{m} A_j,$$

$$A_i = -\frac{\partial \Delta\left(\rho\right)}{\partial \rho_i}\bigg|_{\rho=0},$$

for $i = 1,...,m$, and $\hat{u}$ is a vector of OLS residual from (5.4).

For our problem of testing (5.12) against (5.13), the LMMPI test is determined by (5.14) with $A = (I_n - D)$, which is given by (3.16), where $D = I_m \otimes E_k$. Substituting A in (5.14), we can express d' as

$$d' = \frac{\hat{u}'(I_n - D)\hat{u}}{\hat{u}'\hat{u}}$$

$$= 1 - \frac{\hat{u}'D\hat{u}}{\hat{u}'\hat{u}}$$

$$= 1 - \frac{\displaystyle\sum_{i=1}^{m}\left(\sum_{j=1}^{k}\hat{u}_{ij}\right)^2}{\displaystyle\sum_{i=1}^{m}\sum_{j=1}^{k}\hat{u}_{ij}^2}. \tag{5.15}$$

It is worth noting that d' is the LBI test of $\overline{H}_0$ in the direction of $\rho_1 = \rho_2 = ... = \rho_m > 0$, (see King and Wu (1990) for a property of the LMMPI test), and that d' test is equivalent to the LM1 test of H_0 against H_a which was introduced by Deaton and Irish (1983). Observe that the form of equation (5.15) is a ratio of quadratic forms in normal variables, and therefore its critical values may be determined by the standard numerical techniques mentioned earlier. Under H_0, the LM1 test statistic has a standard normal asymptotic distribution while the LM2 test statistic is asymptotically distributed as $\chi_{(1)}^2$. The next section of this chapter will compare the powers of these tests with that of the PE and the existing tests.

5.3. AN EMPIRICAL COMPARISON OF SIZES AND POWERS

The aim of this section is to explore the small-sample power properties of the $s(\rho_1)$ test of H_0 against H_a, when the intra-cluster or equicorrelation coefficient is present in the two-stage linear regression model. The power functions of the $s(\rho_1)$ tests is evaluated in the context of model (5.4) and (5.5) and is compared with that of the DW (d), King's (1981) modified DW (d^*), the one-sided and two-sided Lagrange multiplier (LM1 and LM2) tests (as calculated and reported by King and Evans, 1986). Exact five percent critical values of the $s(\rho_1)$ test and the associated values of the power at $\rho = 0.1, 0.2,$ 0.3, 0.4 and 0.5 are computed using a modified version of the FQUAD subroutine of Koerts and Abrahamse (1969), with maximum integration and truncation errors of 10^{-6}.

The X matrices used by King and Evans are based on cross-sectional Australian census data, which is classified into 64 demographic groups according to 8 categories of sex/marital status and 8 categories of age (see Williams and Sams (1981) for detailed discussion of the data). For these categories the regressors are population, number of households, and headship ratios of households. The first three data sets consisted of a constant dummy and all combinations of two of these three regressors for 1961, and when extra observations were necessary, 1976 data sets were used. The other two design matrices are composed of a constant and all three regressors for 1961 and 1976, and for 1966 and 1971, respectively. In each case, 32, 64 and 96 observations are used, each with equal cluster sizes of $k = 2, 4, 8$ and $\frac{n}{2}$ (i.e., the cluster size equivalent to half of the sample size). It is also assumed that the data used in the analysis is ordered, so that observations from the same cluster were together. The values of the variables used in this study are presented in tables 5.1, 5.2 and 5.3 of appendix 5A for ready reference and further research.

For a cluster size of $\frac{n}{2}$ our results appear to show that the critical region of the $s(\rho_1)$ test is approximately invariant to the choice of ρ_1 value, and hence it is approximately UMPI. To investigate this possibility, one can re-examine the $s(\rho_1)$ test.

Using (5.8) the critical region of the $s(\rho_1)$ test can also be written as

$$\frac{\sum_{i=1}^{n-p} \lambda_i \xi_i^2}{\sum_{i=1}^{n-p} \xi_i^2} < c_\alpha, \tag{5.16}$$

where in this case the λ_i's are the eigenvalues of the matrix

$$A = \Delta_1^{-1} - \Delta_1^{-1} X \left(X' \Delta_1^{-1} X \right)^{-1} X' \Delta_1^{-1},$$

or equivalently of

$$\left[I_n - \left(\Delta_1^{-\frac{1}{2}} \right)' X \left(X' \Delta_1^{-1} X \right)^{-1} X' \left(\Delta_1^{-\frac{1}{2}} \right) \right] \Delta_1^{-1}.$$

By using King's (1987a, p.23) Corollary and a knowledge of the eigenvalues of Δ_1^{-1} from lemma 1 of appendix 3A, we have the following relationships

$$\begin{cases} \lambda_i = \dfrac{1}{\left\{ 1 + (k-1)\rho_1 \right\}}, \text{ for } i = 1,...,m-p \\[4mm] \dfrac{1}{\left\{ 1 + (k-1)\rho_1 \right\}} \leq \lambda_i \leq \dfrac{1}{1-\rho_1}, \text{ for } i = m-p+1,...,m \\[4mm] \lambda_i = \dfrac{1}{1-\rho_1}, \text{ for } i = m+1,...,n-p. \end{cases} \tag{5.17}$$

If $m > p$, then (5.16) can be expressed as

$$\frac{\left\{\frac{1}{1+(k-1)\rho_1}\right\}\sum_{i=1}^{m-p}\xi_i^2 + \sum_{i=m-p+1}^{m}\lambda_i\xi_i^2 + \frac{1}{1-\rho_1}\sum_{i=m+1}^{n-p}\xi_i^2}{\sum_{i=1}^{n-p}\xi_i^2} < c_\alpha. \qquad (5.18)$$

In this case, when m is small and n is large, the last term of the numerator of (5.18) dominates and hence, $(1-\rho_1)s(\rho_1)$ approximately does not depend on ρ_1. This implies that the $s(\rho_1)$ test for $m > p$ is approximately UMPI.

If $m \le p$, then (5.18) becomes

$$\frac{\sum_{i=1}^{n-p}\lambda_i\xi_i^2 + \frac{1}{1-\rho_1}\sum_{i=m+1}^{n-p}\xi_i^2}{\sum_{i=1}^{n-p}\xi_i^2} < c_\alpha, \qquad (5.19)$$

where the eigenvalues of the numerator of the first term in (5.19) are given by (5.17), i.e. $\dfrac{1}{\left[1+(k-1)\rho_1\right]} \le \lambda_i \le \dfrac{1}{1-\rho_1}$, for $i=1,\dots,m$.

Exact values of these eigenvalues for $m = 2$ are computed by using the numerical algorithm TRED2 from Martin et al. (1968). For each of King and Evans' data sets, it is found that all of the eigenvalues are, $\dfrac{1}{1-\rho_1}$ except for one, which is very small and is approximately equal to $\dfrac{1}{\left[1+(k-1)\rho_1\right]}$. If we ignore the smallest eigenvalue, then the critical regions (5.19) become

$$s^* = \frac{\sum_{i=2}^{n-p}\xi_i^2}{\sum_{i=1}^{n-p}\xi_i^2} < c_\alpha(1-\rho_1). \qquad (5.20)$$

From (5.20) it is clear that the test is approximately independent of the values of ρ_1, and hence it is almost invariant to the choice of ρ_1. Therefore the test based on $s(\rho_1)$ for $m \leq p$ is an approximate UMPI. This also includes the case of cluster size, $k = \dfrac{n}{2}$ or equivalently, $m = 2$, which makes this approximation stronger. This can also be verified by looking at the last clusters of the tables 5.7 to 5.15 in appendix 5B, where the power of the $s(\rho_1)$ test for the cluster size $\dfrac{n}{2}$ is unchanged up to the four decimal places. Hence we have the following proposition for the $s(\rho_1)$ test.

Proposition 5.1. The $s(\rho_1)$ test is approximately UMPI for all values of p and m.

It is worth noting that in (5.20), $s*$ can be expressed in terms of the standard F distribution as follows

$$\Pr\left[s* < c_\alpha\left(1-\rho_1\right)\right] = \Pr\left[\frac{1}{s*} > \frac{1}{c_\alpha^*\left(1-\rho_1\right)}\right]$$

$$= \Pr\left[\frac{\xi_1^2}{\displaystyle\sum_{i=2}^{n-p}\xi_i^2} < \left(\frac{1}{c_\alpha^*\left(1-\rho_1\right)} - 1\right)\right]$$

$$= \Pr\left[\frac{\chi_{(1)}^2}{\chi_{(n-p-1)}^2 / \left(n-p-1\right)} > \left(\frac{1}{c_\alpha^*\left(1-\rho_1\right)} - 1\right)\left(n-p-1\right)\right]$$

$$= \Pr\left[F\left(\vartheta_1, \vartheta_2\right) > F_\alpha\right],$$

where

$$F_\alpha = \left(\frac{1}{c_\alpha^* \left(1 - \rho_1 \right)} - 1 \right) \left(n - p - 1 \right),$$

$$\vartheta_1 = 1, \vartheta_2 = n - p - 1$$

and $F(\vartheta_1, \vartheta_2)$ denotes the F distribution with ϑ_1 and ϑ_2 degrees of freedom. If F_α is the $100(1-\alpha)$ percentile of the F distribution (obtainable from standard Tables), then the approximate critical value, c_α^* can be obtained by the following equation

$$c_\alpha^* = \frac{1}{\left(1 - \rho_1 \right) \left(\dfrac{F_\alpha}{\left(n - p - 1 \right)} + 1 \right)}.$$

5.4. APPROXIMATION OF   AND SIZES USING
F DISTRIBUTION

A comparison of the approximate critical values, c_α^*, and the exact critical values, c_α, for different values of n, $k = \dfrac{n}{2}$ (and/or $m=2$) at five percent significant level is given in tables 5.4, 5.5 and 5.6, of appendix 5B. This is done using cross-sectional Australian census data of appendix 5A for the years (1961 and 1976), (1966 and 1971) and artificially generated log-normal variable data. The last set of data has been used as a proxy for cross-sectional data in many empirical studies, with examples which include Goldfeld and Quandt (1965), Harvey and Phillips (1974), Harrison and McCabe (1979), Prakash (1979), Harrison (1980) and Buse (1980), among others. The results reported in the last column of the tables 5.4 to 5.6 are the estimated sizes of the POI test, which are based on approximate critical values obtained using the standard F distribution. The exact values of the F distribution with ϑ_1, and ϑ_2 degrees of freedom are obtained by using the CDFFC subroutine of GAUSS system. It is worth noting that the results of the exact critical values, c_α and the approximate critical values, c_α^* are almost similar and can be seen in tables 5.4 and 5.5. This is because the same households have been used in the samples with different time periods, with the change in data set and the results in table 5.6 because the data set of a different nature has been considered.

The results of the tests reveal that in all cases the estimated sizes of the $s^*(\rho_1)$ test are less than the nominal level and greater than half of the nominal level, 0.05, even in small samples, except for $n=32$, $\rho_1=0.1, 0.2$ and for $n=64$, $\rho_1=0.1$. It can be noticed in tables 5.4 to 5.6 that there is a tendency for the estimated sizes of the $s^*(\rho_1)$ test to become closer to the nominal sizes when the sample size and the value of ρ_1 increases. This is one of the desirable properties of the POI test, as it always has sizes over the null parameter space closer to the nominal significance level 0.05. The POI test has the same property for large and moderate sample sizes. An advantage of this is, that applied econometricians can use the standard F distribution to calculate approximate critical values of the requisite POI test.

In the rest of this section has been divided into two parts. The first part reports the power calculations for the five $s(\rho_1)$ tests with $\rho_1=0.1, 0.2, 0.3, 0.4$ and 0.5, respectively. After comparing the power performance of these tests, the test whose power is the closest to the power envelope (PE) is chosen. The second section compares the power performance of a selected $s(\rho_1)$ test with that of the LM1, LM2, and d^* tests, from the King and Evans study and the PE.

Power Comparisons of S(ρ_1) Tests

Selected results of the calculations of the power of the first part are given in tables 5.7 to 5.15 of appendix 5B, with the five percent critical values, c_α, provided in column one. In general, under H_a, the power function of all tests increases with the increase in sample size, cluster size and the value of ρ.

As noted above, for a cluster of size two, all of the $s(\rho_1)$ tests are approximately UMPI tests, as they are close to the PE. For a cluster size of $\frac{n}{2}$, they are almost UMPI tests as their power is very close to the PE, (as far as these design matrices are concerned), since the critical regions for this particular clusters size is almost invariant to the choice of ρ_1 values. For example, consider all of the $s(\rho_1)$ tests, for $\rho_1 = 0.1$ to 0.5, for $n = 32$ and cluster of size two. It is found that the power curves of all tests are approximately identical to the PE for all values of $\rho \leq 0.5$. The maximum power difference of these tests lies in the interval of 0.0001 to 0.0007, which is negligible for lower values of ρ and/or large sample sizes.

For a comparison of higher cluster of sizes 4, 8 and 16, the $s(.1)$, $s(.2)$, $s(.3)$, $s(.4)$ and $s(.5)$ tests are MPI in the neighbourhood of $\rho= 0.1, 0.2, 0.3, 0.4$ and 0.5, respectively. The choice between the $s(.1)$ to $s(.5)$ tests depends upon the researcher's requirement. If a researcher is interested in detecting larger values of ρ in smaller sample sizes, then perhaps the $s(.3)$, $s(.4)$ or $s(.5)$ tests could be favoured. If one is however interested in detecting smaller values of ρ, then the $s(.1)$ or $s(.2)$ tests could be more suitable. For routine use, the utilization of the $s(.3)$ test is preferred, as under H_a the value of $\rho_1 = 0.3$ seems to be more practicable, with the calculations of its powers and sizes is the closest to the PE and the nominal size of 0.05, as compared to other $s(\rho_1)$ tests under consideration. With the increase in the sample size this maximum difference between the PE and the power of $s(.3)$ tests reduces, and increases slightly with the higher cluster sizes.

Another procedure for comparing the powers of such tests is given by Cox and Hinkley (1974, p. 102). Cox and Hinkley suggest to choose a test which maximizes some weighted average of the powers. The problem with this procedure is how to determine the appropriate weighted average, as the selected weights may vary from one person to another. To overcome this problem, Brooks' (1993) suggests taking a simple average of the powers. Taking on Brooks' suggestions, we consider a simple average of the powers over a set of ρ values (i.e., $\rho=0.1,0.2,0.3,0.4,0.5$) and then compare these between tests.

The average power of these tests is given in the last column of tables 5.7, 5.10 and 5.11 for the design matrices of (1961 and 1976), whereas the results of tables 5.8, 5.11 and 5.14 are obtained from the data of (1966 and 1971). Tables 5.9, 5.12, and 5.15 show the results for the artificially generated data, which is a procedure that also favours our selection of the $s(.3)$ test for moderate sample sizes. The exceptional cases are for $n=32$, $k=2, 4$, $n=64$, $k=16$, and $n=96$, $k=4, 8$, where the average of the power of the $s(.4)$ and $s(.2)$ tests, lead to that of the $s(.3)$ test by 0.0001. This difference in average power is very small, and therefore it is relatively unimportant for practical purposes. The PE calculated for different values of n and ρ are given in tables 5.16 to 5.21 for the first two design matrices of Australian cross-sectional data. The power patterns of the third data matrix are quite similar to the previous ones, and therefore their results are omitted here. Hence in the next part of this investigation, the power of the $s(.3)$ test will be compared with that of the LM1, LM2, d and d^* tests. Here some of the power calculations are reproduced from King and Evans' (1986) table, so that the second part of our investigation should be self-contained.

5.5. POWER COMPARISONS OF $S(.3)$, LM1, LM2, D^* AND D TESTS WITH PE

The numerical results for the powers of the $s(.3)$, LM1, LM2, d^* and d tests are tabulated in tables 5.16 to 5.21, for different sample sizes, i.e. $n = 32$, 64 and 96. These results demonstrate that the powers of all tests increase as n, k and ρ increase, while other variables remain the same. As expected, the $s(.3)$ test is always the most powerful of the four tests for almost all values of ρ, followed by the LM1, LM2, d^* and d tests. The power difference between the $s(.3)$ and LM1 tests for smaller cluster sizes is very small however. For example, consider the cluster size of 2 in tables 5.16 to 5.21. It is found that the power of both the tests is approximately identical; with the increase in the sample sizes as both the power curves merge. Hence it is confirmed that the LM1 test is a good approximation to the $s(.3)$ test for such a special case. Therefore, the LM1 test is also an approximately UMPI test as is the case with the $s(.3)$ test, at least for these design matrices.

For the higher cluster sizes, the power of the $s(.3)$ test is superior to that of the LM1 test for all values of $\rho \geq 0.2$. This can be verified by considering $n = 32$, $k = 4$, where the maximum power difference between the LM1 and $s(.3)$ tests varies from 0.001 to 0.017, and for $k = 8$, it varies from 0.006 to 0.028. As the sample size, cluster size and the value of ρ increase, this maximum difference between the powers becomes smaller and smaller, and the average powers of the $s(.3)$ and LM1 tests gets closer and closer. For example, consider the case of $n = 96$, $k = 8$. For $0.2 \leq \rho \leq 0.5$, the power difference between the two tests varies from 0.003 to 0.001, with the average power difference is 0.002. This particular case the power of the $s(.3)$ test is identical to that of the PE, which on the basis of these facts, it may be concluded that the POI test is marginally better than the LM1 test at least for small and moderate sample sizes.

As King and Evans noted, the Durbin-Watson (d) test is the least powerful of all the tests under comparison, though it still has reasonable power in the larger samples and the smaller cluster sizes. The advantages of using the d test over other tests are, firstly, it has reasonable small sample power properties when the cluster structure is unavailable, and secondly, its computations can easily be undertaken by standard regression packages, with the availability of existing bounds in the econometrics literature. Further details of the comparisons of these existing tests can be found in King and Evans (1986).

5.6. CONCLUDING REMARKS

This chapter considered the two-stage linear regression model with equicorrelated disturbances, and derives POI tests for the hypothesis that the equicorrelation coefficient, ρ, has any given value. The power of the POI test was compared with that of the existing tests and the PE, with the results in Appendix 5B suggesting that the POI test is marginally better than the LM1 test for small and moderate sample sizes (at least for the data sets used in this experiment). It was found that both the tests are approximately UMPI for some selected cluster sizes, and that for all values of m and p, the POI test is approximately UMPI. Further, the critical values of the POI test can also be obtainable from the standard F distribution for some selected cluster sizes. The approximate critical values using the F distribution was calculated, with 87% of the cases the estimated sizes based on these critical values being greater than half of the nominal sizes. A LMMPI test for testing the hypothesis that ρ has different values for each cluster has been derived, and it was found that the LMMPI test is equivalent to the LM1 test. In the next chapter a testing procedure for the multiple cluster effects is considered, and the question of how close the LMMPI test is to that of the PE in the case of 3SLR model, is discussed.

APPENDIX 5A: THE DATA USED IN THIS CHAPTER

The symbols used here have already been described in the chapter, where $i=1,...,m$, refers to the number of clusters and $j=1,...,k$, the number of observations in the ith cluster.

Table 5.1. Australian cross-sectional data for 1961 and 1976 for $n=96$,
$m=12$ and $k=8$

I	J	x_{ij2}	x_{ij3}	x_{ij4}	i	j	x_{ij2}	x_{ij3}	x_{ij4}
1	1	4.12100	0.02555	0.0062	4	1	0.00020	0.00001	0.0500
1	2	2.63500	0.11010	0.0418	4	2	0.00149	0.00020	0.1342
1	3	1.87800	0.24480	0.1304	4	3	0.01317	0.00481	0.3652
1	4	0.89820	0.21520	0.2396	4	4	0.04521	0.02648	0.5857
1	5	0.63030	0.21200	0.3364	4	5	0.11530	0.07438	0.6452
1	6	0.24350	0.09219	0.3786	4	6	0.09287	0.05991	0.6451
1	7	0.20460	0.07880	0.3852	4	7	0.12980	0.07857	0.6054
1	8	0.38840	0.15510	0.3994	4	8	0.76280	0.39370	0.5161
2	1	0.03885	0.01566	0.4031	5	1	3.69400	0.01921	0.0052
2	2	0.97860	0.67680	0.6916	5	2	1.32600	0.04987	0.0376
2	3	5.35100	4.53400	0.8473	5	3	0.65880	0.05916	0.0898
2	4	6.34100	5.73600	0.9046	5	4	0.46330	0.07904	0.1706
2	5	5.42600	4.95500	0.9132	5	5	0.48050	0.13710	0.2853
2	6	2.00100	1.81800	0.9084	5	6	0.23300	0.08672	0.3722
2	7	1.54300	1.38800	0.8996	5	7	0.22960	0.09442	0.4113
2	8	2.64900	2.32200	0.8766	5	8	0.64360	0.27550	0.4281
3	1	0.00009	0.00000	0.0500	6	1	0.27610	0.00218	0.0079
3	2	0.00180	0.00016	0.0889	6	2	2.02800	0.02961	0.0416
3	3	0.04473	0.01209	0.2703	6	3	5.89300	0.15150	0.0257
3	4	0.10790	0.04122	0.3819	6	4	6.48800	0.23940	0.0369
3	5	0.11680	0.04937	0.4226	6	5	4.92400	0.23440	0.0476
3	6	0.04282	0.01870	0.4367	6	6	1.61200	0.08206	0.0509
3	7	0.03126	0.01354	0.4331	6	7	1.29000	0.07017	0.0544
3	8	0.04093	0.01796	0.4388	6	8	1.87800	0.13630	0.0726
7	1	0.00030	0.00003	0.1000	10	1	0.06212	0.04344	0.6671
7	2	0.00521	0.00068	0.1305	10	2	1.83700	1.55700	0.8475
7	3	0.06138	0.01899	0.3094	10	3	8.19900	7.53000	0.9184
7	4	0.12250	0.05923	0.4834	10	4	6.92300	6.56100	0.9477
7	5	0.12400	0.07017	0.5660	10	5	6.68500	6.34100	0.9486
7	6	0.04237	0.02350	0.5546	10	6	2.65000	2.49400	0.9411
7	7	0.03383	0.01829	0.5406	10	7	2.28300	2.12000	0.9290
7	8	0.04378	0.02294	0.5240	10	8	3.70000	3.32600	0.8989
8	1	0.00071	0.00009	0.1268	11	1	0.00055	0.00012	0.2182
8	2	0.00498	0.00148	0.2972	11	2	0.02006	0.00830	0.4138
8	3	0.03950	0.02376	0.6016	11	3	0.22330	0.13280	0.5949
8	4	0.15570	0.11420	0.7334	11	4	0.21490	0.13580	0.6321
8	5	0.43840	0.32400	0.7391	11	5	0.23890	0.15200	0.6363
8	6	0.36430	0.25490	0.6996	11	6	0.09308	0.05882	0.6319
8	7	0.54610	0.36350	0.6656	11	7	0.07435	0.04426	0.5953
8	8	2.53700	1.40600	0.5544	11	8	0.09513	0.05479	0.5759
9	1	6.15200	0.13720	0.0223	12	1	0.00061	0.00012	0.1967
9	2	3.72800	0.60800	0.1631	12	2	0.00257	0.00090	0.3502
9	3	2.08600	0.68540	0.3286	12	3	0.01539	0.00876	0.5692
9	4	0.71950	0.28880	0.4014	12	4	0.03801	0.02700	0.7103
9	5	0.65550	0.31160	0.4753	12	5	0.12640	0.09649	0.7634
9	6	0.23960	0.12600	0.5260	12	6	0.10500	0.07670	0.7308
9	7	0.20910	0.11500	0.5502	12	7	0.14970	0.10760	0.7191
9	8	0.41130	0.20610	0.5011	12	8	0.85720	0.50310	0.5869

Table 5.2. Australian cross-sectional data for 1966 and 1971 for n=96, m=2, k=48, i.e. i=1,2 and j=1,...,48

i	J	x_{ij2}	x_{ij3}	x_{ij4}	i	J	x_{ij2}	x_{ij3}	x_{ij4}
1	1	5.30000	0.03816	0.0072	1	37	0.41670	0.13410	0.3217
1	2	3.05300	0.16760	0.0549	1	38	0.22290	0.09266	0.4157
1	3	1.70300	0.27380	0.1608	1	39	0.21990	0.10010	0.4551
1	4	0.92600	0.25690	0.2774	1	40	0.67400	0.31150	0.4621
1	5	0.62020	0.23440	0.3779	1	41	0.41560	0.00424	0.0102
1	6	0.25580	0.10920	0.4269	1	42	2.47000	0.04841	0.0196
1	7	0.20670	0.09189	0.4445	1	43	6.16300	0.19410	0.0315
1	8	0.39730	0.17040	0.4290	1	44	6.73800	0.29240	0.0434
1	9	0.06853	0.03761	0.5488	1	45	5.48500	0.26440	0.0482
1	10	1.30500	0.99830	0.7652	1	46	1.96700	0.09618	0.0489
1	11	5.64200	4.93100	0.8739	1	47	1.38000	0.06762	0.0490
1	12	6.85500	6.32300	0.9224	1	48	2.02800	0.11660	0.0575
1	13	5.80200	5.39500	0.9298					
1	14	2.35100	2.17000	0.9232	2	1	0.00083	0.00003	0.0361
1	15	1.78000	1.62600	0.9134	2	2	0.00733	0.00163	0.2224
1	16	2.87000	2.54500	0.8867	2	3	0.06373	0.02638	0.4139
1	17	0.00102	0.00007	0.0686	2	4	0.12950	0.07305	0.5642
1	18	0.00373	0.00054	0.1448	2	5	0.14500	0.09018	0.6220
1	19	0.04243	0.01470	0.3464	2	6	0.05808	0.03590	0.6181
1	20	0.11010	0.04925	0.4474	2	7	0.04308	0.02592	0.6017
1	21	0.12740	0.06448	0.5061	2	8	0.06390	0.03633	0.5686
1	22	0.05360	0.02643	0.4931	2	9	0.00151	0.00015	0.0993
1	23	0.03912	0.02021	0.5166	2	10	0.00684	0.00230	0.3363
1	24	0.05147	0.02452	0.4764	2	11	0.03992	0.02643	0.6621
1	25	0.00100	0.00005	0.0500	2	12	0.16050	0.12990	0.8093
1	26	0.00335	0.00040	0.1194	2	13	0.47810	0.37260	0.7794
1	27	0.01274	0.00525	0.4121	2	14	0.41990	0.30570	0.7280
1	28	0.04883	0.03092	0.6332	2	15	0.55240	0.38200	0.6915
1	29	0.11860	0.08314	0.7012	2	16	2.97200	1.70500	0.5738
1	30	0.10050	0.06866	0.6833	2	17	5.58500	0.06143	0.0110
1	31	0.13000	0.08465	0.6514	2	18	3.56700	0.31710	0.0889
1	32	0.80640	0.42790	0.5306	2	19	1.80600	0.39720	0.2199
1	33	4.69900	0.03477	0.0074	2	20	0.82120	0.26850	0.3269
1	34	1.68200	0.10160	0.0604	2	21	0.63840	0.26860	0.4208
1	35	0.66970	0.09041	0.1350	2	22	0.24850	0.12100	0.4869
1	36	0.41280	0.08611	0.2086	2	23	0.20490	0.10350	0.5051
2	24	0.39510	0.19180	0.4855	2	37	0.17340	0.09697	0.5593
2	25	0.08158	0.04674	0.5729	2	38	0.07357	0.04250	0.5777
2	26	2.00000	1.62400	0.8120	2	39	0.05676	0.03222	0.5677
2	27	7.02100	6.31100	0.8989	2	40	0.07699	0.04222	0.5484
2	28	6.88000	6.43100	0.9347	2	41	0.00033	0.00009	0.2727
2	29	6.37700	5.99400	0.9399	2	42	0.00288	0.00094	0.3264
2	30	2.58500	2.41800	0.9354	2	43	0.01566	0.00844	0.5390
2	31	2.03200	1.88100	0.9257	2	44	0.05032	0.03571	0.7097
2	32	3.15600	2.82800	0.8959	2	45	0.13350	0.10110	0.7575
2	33	0.00000	0.00000	0.0000	2	46	0.10810	0.07992	0.7394
2	34	0.01401	0.00325	0.2320	2	47	0.14400	0.10200	0.7086
2	35	0.08930	0.03885	0.4351	2	48	0.84040	0.48120	0.5725

Table 5.3. Artificially generated log-normal variable data for $n=96$, $m=3$, $k=32$, i.e. $i=1,2,3$, $j=1,...,32$

I	j	x_{ij2}	i	j	x_{ij2}	i	j	x_{ij2}
1	1	1.33369	2	1	1.31581	3	1	0.96271
1	2	1.15899	2	2	1.41961	3	2	0.97292
1	3	1.34683	2	3	1.37922	3	3	0.81020
1	4	0.82016	2	4	1.29287	3	4	0.88585
1	5	1.21295	2	5	1.06239	3	5	1.11333
1	6	1.03895	2	6	1.63080	3	6	0.87408
1	7	1.13887	2	7	1.06660	3	7	1.39801
1	8	0.51691	2	8	1.08967	3	8	1.05341
1	9	1.38710	2	9	1.11439	3	9	1.30407
1	10	1.11005	2	10	0.74284	3	10	1.25518
1	11	0.31322	2	11	1.67184	3	11	1.32082
1	12	1.07261	2	12	1.61571	3	12	0.32865
1	13	0.64652	2	13	0.97981	3	13	0.49337
1	14	1.09548	2	14	1.07496	3	14	0.90605
1	15	1.35041	2	15	1.02444	3	15	1.22240
1	16	1.49629	2	16	1.56426	3	16	0.73985
1	17	1.36745	2	17	0.26251	3	17	0.64744
1	18	1.45277	2	18	1.28764	3	18	0.80285
1	19	1.29114	2	19	1.21660	3	19	0.79286
1	20	1.43091	2	20	0.65502	3	20	1.08637
1	21	0.30469	2	21	1.69976	3	21	1.48146
1	22	-0.21046	2	22	1.12795	3	22	1.34389
1	23	1.03297	2	23	1.46930	3	23	0.66093
1	24	1.10924	2	24	1.06172	3	24	1.26205
1	25	1.51238	2	25	1.38130	3	25	0.95833
1	26	1.17286	2	26	1.25050	3	26	1.42739
1	27	0.66592	2	27	-0.41641	3	27	1.22635
1	28	1.42282	2	28	0.95260	3	28	1.41553
1	29	1.34428	2	29	1.01722	3	29	1.09891
1	30	1.20788	2	30	0.70807	3	30	1.14698
1	31	1.16500	2	31	1.48471	3	31	0.81043
1	32	0.80592	2	32	1.46623	3	32	1.45789

APPENDIX 5B: TABULATION OF RESULTS
OF THE COMPARATIVE STUDY

Table 5.4. Comparison of an exact and approximate critical values of $s(\rho_1)$ test for cluster of size $\frac{n}{2}$, $p = 4$, using cross-sectional data for 1961 and 1976

POI Test	ρ_1	Exact critical values, c_α, at $\alpha=.05$	Approximate critical values, c_α^*	Estimated sizes based on c_α^*
$s(\rho_1)$		n=32, F(1,27)=4.2097		
	0.1	1.03417	0.96124	.005
	0.2	1.13137	1.08139	.018
	0.3	1.27380	1.23588	.027
	0.4	1.47252	1.44186	.034
	0.5	1.75592	1.73023	.039
$s(\rho_1)$		n=64, F(1,59)=4.003		
	0.1	1.05750	1.04051	.024
	0.2	1.18040	1.17058	.036
	0.3	1.34470	1.33780	.041
	0.4	1.56608	1.56077	.044
	0.5	1.87723	1.87293	.046
$s(\rho_1)$		n=96, F(1,91)=3.9454		
	0.1	1.07339	1.06494	.030
	0.2	1.20276	1.19806	.040
	0.3	1.37246	1.36921	.044
	0.4	1.59990	1.59741	.046
	0.5	1.91890	1.91689	.047

M. Ishaq Bhatti

Table 5.5. Comparison of an exact and approximate critical values of $s(\rho_1)$ test for cluster of size n/2, p = 4, using cross-sectional data for 1966 and 1971

POI Test	ρ_1	Exact critical values, c_α, at α=.05	Approximate critical values, c_α^*	Estimated sizes based on c_α^*
		n=32, F(1,27)=4.2097		
$s(\rho_1)$	0.1	1.03296	0.96124	.005
	0.2	1.13024	1.08139	.018
	0.3	1.27291	1.23588	.028
	0.4	1.47167	1.44186	.034
	0.5	1.75518	1.73023	.039
		n=64, F(1,59)=4.003		
$s(\rho_1)$	0.1	1.05755	1.04051	.024
	0.2	1.18040	1.17058	.036
	0.3	1.34470	1.33780	.041
	0.4	1.56608	1.56077	.044
	0.5	1.87724	1.87293	.046
		n=96, F(1,91)=3.9454		
$s(\rho_1)$	0.1	1.07293	1.06494	.031
	0.2	1.20248	1.19806	.040
	0.3	1.37226	1.36921	.044
	0.4	1.59974	1.59741	.046
	0.5	1.91877	1.91689	.047

Table 5.6. Comparison of an exact and approximate critical values of $s(\rho_1)$ test for cluster of size $\dfrac{n}{2}$, $p = 2$, using artificially generated log-normal regressor

POI Test	ρ_1	Exact critical values, c_α, at $\alpha=.05$	Approximate critical values, c_α^*	Estimated sizes based on c_α^*
	n=32, F(1,27)=4.2097			
$s(\rho_1)$	0.1	1.021471	0.96124	.009
	0.2	1.123943	1.08139	.022
	0.3	1.271409	1.23588	.029
	0.4	1.474579	1.44186	.033
	0.5	1.762715	1.73023	.036
	n=64, F(1,59)=4.003			
$s(\rho_1)$	0.1	1.057875	1.04051	.023
	0.2	1.181720	1.17058	.034
	0.3	1.134672	1.33780	.039
	0.4	1.568771	1.56077	.041
	0.5	1.880729	1.87293	.043
	n=96, F(1,91)=3.9454			
$s(\rho_1)$	0.1	1.073136	1.06494	.030
	0.2	1.203116	1.19806	.039
	0.3	1.373191	1.36921	.042
	0.4	1.600954	1.59741	.044
	0.5	1.920328	1.91689	.047

M. Ishaq Bhatti

Table 5.7. Powers of the POI-tests for the nx4 design matrix of cross-sectional data for 1961 and 1976, n=32, p=4 and α=.05

Tests	c_α	ρ=0.1	0.2	0.3	0.4	0.5	Average Power
Cluster size 2							
s(.1)	0.97890	.0989	.1783	.2943	.4455	.6180	.3270
s(.2)	0.97657	.0989	.1783	.2944	.4458	.6186	.3272
s(.3)	0.99450	.0989	.1783	.2944	.4460	.6191	.3273
s(.4)	1.03719	.0989	.1782	.2944	.4461	.6196	.3274
s(.5)	1.11435	.0988	.1780	.2941	.4460	.6198	.3273
Cluster size 4							
s(.1)	0.98451	.1397	.2720	.4271	.5831	.7230	.4290
s(.2)	1.01296	.1395	.2724	.4290	.5868	.7282	.4312
s(.3)	1.07863	.1390	.2721	.4295	.5889	.7316	.4322
s(.4)	1.18630	.1383	.2710	.4290	.5895	.7335	.4323
s(.5)	1.35232	.1373	.2692	.4274	.5889	.7341	.4314
Cluster size 8							
s(.1)	1.00513	.1640	.3016	.4375	.5626	.6740	.4279
s(.2)	1.06902	.1637	.3026	.4410	.5684	.6812	.4314
s(.3)	1.17651	.1627	.3021	.4418	.5705	.6843	.4323
s(.4)	1.33503	.1615	.3009	.4413	.5710	.6854	.4320
s(.5)	1.56785	.1604	.2995	.4404	.5707	.6857	.4313
Cluster size 16							
s(.1)	1.03417	.1638	.2738	.3702	.4551	.5318	.3589
s(.2)	1.13137	.1638	.2738	.3702	.4551	.5318	.3589
s(.3)	1.27380	.1638	.2738	.3702	.4551	.5318	.3589
s(.4)	1.47252	.1638	.2738	.3702	.4551	.5318	.3589
s(5)	1.75592	.1638	.2738	.3702	.4551	.5318	.3589

Table 5.8. Powers of the POI-tests for the nx4 design matrix of cross-sectional data for 1966 and 1971, n=32, p=4 and α=.05

Tests	c_α	ρ=0.1	0.2	0.3	0.4	0.5	Average Power
Cluster size 2							
s(.1)	0.97901	.0990	.1785	.2947	.4461	.6186	.3274
s(.2)	0.97683	.0990	.1785	.2947	.4463	.6192	.3275
s(.3)	0.99496	.0990	.1785	.2948	.4465	.6197	.3277
s(.4)	1.03793	.0989	.1784	.2947	.4466	.6203	.3279
s(.5)	1.11546	.0988	.1782	.2945	.4465	.6203	.3277
Cluster size 4							
s(.1)	0.98297	.1402	.2738	.4306	.5883	.7292	.4324
s(.2)	1.01008	.1400	.2742	.4326	.5921	.7343	.4346
s(.3)	1.07452	.1395	.2738	.4331	.5941	.7376	.4356
s(.4)	1.18106	.1388	.2728	.4326	.5947	.7393	.4356
s(.5)	1.34601	.1379	.2712	.4311	.5941	.7399	.4348
Cluster size 8							
s(.1)	1.00358	.1681	.3084	.4450	.5691	.6786	.4338
s(.2)	1.06773	.1676	.3094	.4484	.5750	.6861	.4373
s(.3)	1.17569	.1665	.3088	.4493	.5773	.6895	.4383
s(.4)	1.33462	.1653	.3076	.4488	.5778	.6908	.4381
s(.5)	1.56777	.1640	.3060	.4478	.5775	.6911	.4373
Cluster size16							
s(.1)	1.03296	.1672	.2792	.3763	.4613	.5378	.3644
s(.2)	1.13024	.1672	.2792	.3763	.4613	.5378	.3644
s(.3)	1.27291	.1672	.2792	.3763	.4613	.5377	.3643
s(.4)	1.47167	.1672	.2792	.3763	.4613	.5378	.3644
s(5)	1.75518	.1672	.2792	.3763	.4613	.5378	.3644

Table 5.9. Powers of the POI-tests for the $n \times 2$ design matrix of the artificially generated log-normal regressor, n=32, p=4 and α=.05

Tests	c_α	ρ=0.1	0.2	0.3	0.4	0.5	Average Power
Cluster size 2							
s(.1)	0.97330	.1009	.1849	.3089	.4705	.6517	.3434
s(.2)	0.96527	.1009	.1849	.3090	.4709	.6525	.3436
s(.3)	0.97711	.1009	.1849	.3091	.4712	.6531	.3438
s(.4)	1.01304	.1008	.1848	.3090	.4713	.6536	.3439
s(.5)	1.08243	.1007	.1846	.3088	.4711	.6538	.3438
Cluster size 4							
s(.1)	0.97022	.1562	.3188	.5046	.6783	.8167	.4949
s(.2)	0.99060	.1561	.3189	.5053	.6794	.8179	.4955
s(.3)	1.05195	.1560	.3188	.5054	.6798	.8184	.4957
s(.4)	1.15833	.1558	.3186	.5053	.6799	.8186	.4956
s(.5)	1.32636	.1556	.3183	.5051	.6798	.8187	.4955
Cluster size 8							
s(.1)	0.98236	.2185	.4109	.5757	.7048	.8024	.5425
s(.2)	1.04442	.2185	.4109	.5758	.7048	.8024	.5425
s(.3)	1.15544	.2185	.4109	.5758	.7048	.8024	.5425
s(.4)	1.31982	.2185	.4109	.5758	.7048	.8024	.5425
s(.5)	1.56046	.2185	.4109	.5758	.7048	.8024	.5425
Cluster size 16							
s(.1)	1.02147	.2296	.3679	.4715	.5540	.6236	.4493
s(.2)	1.12394	.2296	.3679	.4715	.5540	.6236	.4493
s(.3)	1.27141	.2296	.3679	.4715	.5540	.6236	.4493
s(.4)	1.47458	.2296	.3679	.4715	.5540	.6236	.4493
s(.5)	1.76272	.2296	.3679	.4715	.5540	.6236	.4493

Table 5.10. Powers of the POI-tests for the *n*x4 design matrix of cross-sectional data for 1961 and 1976, *n*=64, *p*=4, and *α*=.05

Tests	c_α	ρ=0.1	0.2	0.3	0.4	0.5	Average Power
Cluster size 2							
s(.1)	0.98578	.1337	.2878	.5048	.7319	.8990	.5114
s(.2)	0.99114	.1337	.2878	.5049	.7321	.8992	.5115
s(.3)	1.01822	.1337	.2878	.5049	.7323	.8995	.5116
s(.4)	1.07265	.1336	.2877	.5049	.7323	.8996	.5116
s(.5)	1.16612	.1336	.2877	.5049	.7322	.8997	.5116
Cluster size 4							
s(.1)	0.98972	.2198	.4845	.7301	.8887	.9647	.6576
s(.2)	1.02571	.2197	.4848	.7311	.8898	.9654	.6582
s(.3)	1.10161	.2193	.4845	.7314	.8904	.9659	.6583
s(.4)	1.22337	.2187	.4838	.7312	.8906	.9661	.6581
s(.5)	1.40963	.2180	.4826	.7305	.8904	.9662	.6575
Cluster size 8							
s(.1)	1.00735	.2968	.5737	.7696	.8860	.9488	.6950
s(.2)	1.07986	.2962	.5745	.7719	.8884	.9506	.6964
s(.3)	1.19791	.2949	.5742	.7724	.8892	.9513	.6964
s(.4)	1.36903	.2934	.5730	.7721	.8894	.9516	.6959
s(.5)	1.61787	.2919	.5716	.7714	.8893	.9517	.6952
Cluster size 16							
s(.1)	1.02967	.3337	.5705	.7247	.8242	.8898	.6686
s(.2)	1.13044	.3341	.5708	.7252	.8248	.8902	.6690
s(.3)	1.27632	.3331	.5707	.7253	.8249	.8902	.6689
s(.4)	1.47822	.3328	.5706	.7253	.8249	.8903	.6688
s(.5)	1.76525	.3327	.5705	.7252	.8249	.8903	.6687
Cluster size 32							
s(.1)	1.05750	.3297	.4841	.5829	.6552	.7130	.5530
s(.2)	1.18040	.3297	.4841	.5829	.6552	.7130	.5530
s(.3)	1.34470	.3297	.4841	.5829	.6552	.7130	.5530
s(.4)	1.56608	.3297	.4841	.5829	.6552	.7130	.5530
s(.5)	1.87723	.3297	.4841	.5829	.6552	.7130	.5530

Table 5.11. Powers of the POI-tests for the *n*x4 design matrix of cross-sectional data for 1966 and 1971, *n*=64, *p*=4 and α=.05

Tests	c_α	ρ=0.1	0.2	0.3	0.4	0.5	Average Power
Cluster size 2							
s(.1)	0.98576	.1337	.2877	.5046	.7317	.8989	.5113
s(.2)	0.99108	.1337	.2877	.5047	.7319	.8991	.5114
s(.3)	1.01812	.1337	.2877	.5048	.7321	.8994	.5115
s(.4)	1.07249	.1336	.2876	.5047	.7322	.8995	.5115
s(.5)	1.16585	.1335	.2874	.5045	.7321	.8996	.5114
Cluster size 4							
s(.1)	0.98940	.2203	.4855	.7314	.8896	.9651	.6589
s(.2)	1.02518	.2201	.4858	.7323	.8907	.9658	.6589
s(.3)	1.10092	.2197	.4856	.7326	.8913	.9663	.6591
s(.4)	1.22252	.2192	.4849	.7324	.8914	.9665	.6589
s(.5)	1.40862	.2184	.4837	.7317	.8913	.9666	.6583
Cluster size 8							
s(.1)	1.00715	.2982	.5755	.7709	.8867	.9491	.6961
s(.2)	1.07972	.2976	.5765	.7732	.8891	.9509	.6975
s(.3)	1.19786	.2963	.5760	.7737	.8900	.9516	.6975
s(.4)	1.36905	.2948	.5748	.7734	.8901	.9520	.6970
s(.5)	1.61794	.2932	.5733	.7727	.8900	.9520	.6962
Cluster size 16							
s(.1)	1.02978	.3329	.5695	.7238	.8236	.8893	.6678
s(.2)	1.13054	.3326	.5698	.7244	.8241	.8897	.6681
s(.3)	1.27641	.3323	.5698	.7244	.8242	.8898	.6681
s(.4)	1.47829	.3321	.5696	.7244	.8242	.8899	.6680
s(.5)	1.76531	.3319	.5695	.7244	.8242	.8899	.6680
Cluster size 32							
s(.1)	1.05755	.3295	.4840	.5827	.6550	.7129	.5528
s(.2)	1.18040	.3295	.4840	.5827	.6550	.7129	.5528
s(.3)	1.34470	.3295	.4840	.5827	.6550	.7129	.5528
s(.4)	1.56608	.3295	.4840	.5827	.6550	.7129	.5528
s(.5)	1.87724	.3295	.4840	.5827	.6550	.7129	.5528

Table 5.12. Powers of the POI-tests for the nx2 design matrix of the artificially generated log-normal regressor, n=64, p=2 and α=.05

Tests	c_α	ρ=0.1	0.2	0.3	0.4	0.5	Average power
Cluster size 2							
s(.1)	0.98283	.1357	.2937	.5166	.7465	.9100	.5204
s(.2)	0.98511	.1354	.2938	.5167	.7468	.9103	.5206
s(.3)	1.00889	.1354	.2937	.5168	.7470	.9105	.5207
s(.4)	1.05962	.1353	.2936	.5167	.7471	.9107	.5207
s(.5)	1.14882	.1352	.2934	.5165	.7470	.9108	.5206
Cluster size 4							
s(.1)	0.98440	.2336	.5187	.7689	.9156	.9771	.6828
s(.2)	1.01766	.2336	.5188	.7691	.9158	.9772	.6829
s(.3)	1.09239	.2335	.5187	.7692	.9158	.9773	.6829
s(.4)	1.21420	.2334	.5186	.7692	.9159	.9773	.6829
s(.5)	1.40184	.2333	.5185	.7691	.9159	.9773	.6828
Cluster size 8							
s(.1)	0.99804	.3408	.6461	.8333	.9283	.9719	.7441
s(.2)	1.06948	.3408	.6461	.8333	.9283	.9719	.7441
s(.3)	1.18875	.3408	.6461	.8333	.9283	.9719	.7441
s(.4)	1.36214	.3408	.6461	.8333	.9283	.9719	.7441
s(.5)	1.61410	.3408	.6461	.8334	.9283	.9719	.7441
Cluster size 16							
s(.1)	1.02404	.3979	.6444	.7850	.8682	.9198	.7231
s(.2)	1.12727	.3979	.6444	.7850	.8682	.9198	.7231
s(.3)	1.27539	.3979	.6444	.7850	.8682	.9198	.7231
s(.4)	1.47933	.3979	.6444	.7850	.8682	.9198	.7231
s(.5)	1.76849	.3979	.6444	.7850	.8682	.9198	.7231
Cluster size 32							
s(.1)	1.05787	.3508	.5058	.6025	.6723	.7278	.5718
s(.2)	1.18172	.3508	.5058	.6025	.6723	.7129	.5718
s(.3)	1.34672	.3508	.5058	.6025	.6723	.7129	.5718
s(.4)	1.56877	.3508	.5058	.6025	.6723	.7129	.5718
s(.5)	1.88073	.3508	.5058	.6025	.6723	.7129	.5718

M. Ishaq Bhatti

Table 5.13. Powers of the POI-tests for the nx4 design matrix of cross-sectional data for 1961 and 1976, n=96, p=4 and α=.05

Tests	c_α	ρ=0.1	0.2	0.3	0.4	0.5	Average power
Cluster size 2							
s(.1)	0.98950	.1647	.3858	.6639	.8808	.9771	.6145
s(.2)	0.99894	.1646	.3859	.6640	.8809	.9772	.6145
s(.3)	1.03081	.1646	.3858	.6640	.8810	.9773	.6145
s(.4)	1.09127	.1646	.3858	.6640	.8810	.9773	.6145
s(.5)	1.19295	.1645	.3856	.6638	.8810	.9773	.6144
Cluster size 4							
s(.1)	0.99500	.2887	.6389	.8779	.9723	.9959	.7547
s(.2)	1.03622	.2886	.6391	.8784	.9726	.9960	.7549
s(.3)	1.11791	.2881	.6387	.8785	.9727	.9960	.7548
s(.4)	1.24677	.2877	.6384	.8784	.9728	.9961	.7547
s(.5)	1.44241	.2870	.6375	.8781	.9728	.9961	.7543
Cluster size 8							
s(.1)	1.01165	.4056	.7466	.9119	.9734	.9931	.8061
s(.2)	1.08858	.4051	.7472	.9127	.9739	.9933	.8064
s(.3)	1.21144	.4040	.7469	.9129	.9742	.9934	.8063
s(.4)	1.38817	.4028	.7462	.9128	.9742	.9935	.8059
s(.5)	1.64415	.4014	.7452	.9125	.9742	.9935	.8054
Cluster size 16							
s(.1)	1.03267	.4741	.7561	.8872	.9477	.9764	.8083
s(.2)	1.13678	.4738	.7564	.8875	.9480	.9765	.8084
s(.3)	1.28555	.4738	.7563	.8876	.9481	.9766	.8085
s(.4)	1.49049	.4731	.7562	.8876	.9481	.9766	.8083
s(.5)	1.78123	.4728	.7561	.8875	.9481	.9766	.8082
Cluster size 48							
s(.1)	1.07339	.3977	.5515	.6428	.7072	.7576	.6114
s(.2)	1.20276	.3977	.5515	.6428	.7072	.7576	.6114
s(.3)	1.37246	.3977	.5515	.6428	.7072	.7576	.6114
s(.4)	1.59990	.3977	.5515	.6428	.7072	.7576	.6114
s(.5)	1.91890	.3977	.5515	.6428	.7072	.7576	.6114

Table 5.14. Powers of the POI-tests for the nx4 design matrix of cross-sectional data for 1966 and 1971, n=96, p=4 and α=.05

Tests	c_α	ρ=0.1	0.2	0.3	0.4	0.5	Average power
Cluster size 2							
s(.1)	0.98948	.1646	.3857	.6637	.8807	.9771	.6144
s(.2)	0.99888	.1646	.3858	.6638	.8808	.9771	.6144
s(.3)	1.03070	.1646	.3857	.6638	.8809	.9772	.6144
s(.4)	1.09110	.1646	.3856	.6638	.8809	.9773	.6143
s(.5)	1.19267	.1645	.3854	.6636	.8809	.9773	.6143
Cluster size 4							
s(.1)	0.99479	.2891	.6396	.8785	.9725	.9960	.7551
s(.2)	1.03587	.2890	.6398	.8790	.9728	.9960	.7553
s(.3)	1.11746	.2886	.6396	.8791	.9730	.9961	.7553
s(.4)	1.24618	.2881	.6391	.8790	.9730	.9961	.7551
s(.5)	1.44169	.2874	.6382	.8786	.9730	.9961	.7547
Cluster size 8							
s(.1)	1.01121	.4080	.7494	.9134	.9740	.9933	.8076
s(.2)	1.08807	.4075	.7500	.9143	.9746	.9935	.8080
s(.3)	1.21090	.4065	.7497	.9144	.9748	.9936	.8078
s(.4)	1.38764	.4052	.7489	.9143	.9748	.9937	.8074
s(.5)	1.64361	.4040	.7481	.9141	.9748	.9937	.8069
Cluster size 16							
s(.1)	1.03208	.4802	.7616	.8905	.9496	.9773	.8118
s(.2)	1.13627	.4799	.7618	.8908	.9498	.9774	.8119
s(.3)	1.28513	.4795	.7618	.8909	.9498	.9775	.8119
s(.4)	1.49013	.4792	.7617	.8909	.9498	.9775	.8118
s(.5)	1.78092	.4792	.7616	.8908	.9498	.9775	.8118
Cluster size 48							
s(.1)	1.07293	.4107	.5636	.6533	.7162	.7652	.6218
s(.2)	1.20248	.4107	.5636	.6533	.7162	.7652	.6218
s(.3)	1.37226	.4107	.5636	.6533	.7162	.7652	.6218
s(.4)	1.59974	.4107	.5636	.6533	.7162	.7652	.6218
s(.5)	1.91877	.4107	.5636	.6533	.7162	.7652	.6218

M. Ishaq Bhatti

Table 5.15. Powers of the POI-tests for the nx2 design matrix of artificially generated log-normal regressor, $n=96$, $p=2$ and $\alpha=.05$

Tests	c_α	$\rho=0.1$	0.2	0.3	0.4	0.5	Average power
Cluster size 2							
s(.1)	0.98743	.1662	.3913	.6728	.8882	.9800	.6197
s(.2)	0.99470	.1662	.3913	.6729	.8883	.9801	.6198
s(.3)	1.02422	.1662	.3913	.6729	.8883	.9801	.6198
s(.4)	1.08204	.1661	.3912	.6728	.8884	.9801	.6197
s(.5)	1.18066	.1660	.3910	.6727	.8883	.9801	.6196
Cluster size 4							
s(.1)	0.99157	.3016	.6646	.8968	.9794	.9974	.7680
s(.2)	1.03103	.3015	.6646	.8969	.9795	.9974	.7680
s(.3)	1.11199	.3015	.6646	.8969	.9795	.9974	.7680
s(.4)	1.24083	.3014	.6645	.8969	.9795	.9974	.7679
s(.5)	1.43734	.3013	.6044	.8969	.9795	.9974	.7679
Cluster size 8							
s(.1)	1.00607	.4430	.7910	.9370	.9835	.9963	.8302
s(.2)	1.08236	.4430	.7910	.9370	.9835	.9963	.8302
s(.3)	1.20591	.4430	.7910	.9370	.9835	.9963	.8302
s(.4)	1.38397	.4430	.7910	.9370	.9835	.9963	.8302
s(.5)	1.64179	.4430	.7910	.9370	.9835	.9963	.8302
Cluster size 16							
s(.1)	1.02952	.5252	.7998	.9131	.9615	.9831	.8365
s(.2)	1.13507	.5252	.7998	.9131	.9615	.9831	.8365
s(.3)	1.28517	.5252	.7998	.9131	.9615	.9831	.8365
s(.4)	1.49133	.5252	.7998	.9131	.9615	.9831	.8365
s(.5)	1.78334	.5252	.7998	.9131	.9615	.9831	.8365
Cluster size 48							
s(.1)	1.07314	.4294	.5807	.6679	.7286	.7758	.6365
s(.2)	1.20312	.4294	.5807	.6679	.7286	.7758	.6365
s(.3)	1.37319	.4294	.5807	.6679	.7286	.7758	.6365
s(.4)	1.60095	.4294	.5807	.6679	.7286	.7758	.6365
s(.5)	1.92033	.4294	.5807	.6679	.7286	.7758	.6365

Table 5.16. Powers of the Durbin-Watson, one- and two-sided Lagrange multiplier and POI-tests for the nx4 design matrix of cross-sectional data for 1961 and 1976, $n=32$, $p=4$ and $\alpha=.05$

Tests	$\rho=0.1$	0.2	0.3	0.4	0.5	Average power
Cluster size 2						
PE	.099	.178	.294	.446	.620	-
s(.3)	.099	.178	.294	.446	.619	.3272
LM1	.099	.178	.294	.445	.617	.3266
LM2	.066	.112	.196	.324	.492	.2380
d^*	.081	.126	.186	.265	.363	.2042
D	.081	.125	.184	.260	.356	.2012
Cluster size 4						
PE	.140	.272	.430	.590	.734	-
s(.3)	.139	.272	.430	.589	.732	.4324
LM1	.139	.271	.423	.577	.715	.4250
LM2	.094	.190	.323	.475	.627	.3418
d^*	.100	.175	.272	.390	.522	.2918
D	.099	.172	.268	.384	.516	.2878
Cluster size 8						
PE	.164	.303	.442	.571	.686	-
s(.3)	.163	.302	.442	.571	.684	.4321
LM1	.163	.296	.427	.547	.656	.4178
LM2	.118	.224	.343	.463	.578	.3452
d^*	.095	.162	.246	.345	.454	.2604
D	.095	.161	.245	.343	.452	.2592

Table 5.17. Powers Of The Durbin-Watson, One- And Two-Sided Lagrange Multiplier And POI-Tests For The *N*x4 Design Matrix Of Cross-Sectional Data For 1966 And 1971, *N*=32, *P*=4 And α=.05

Tests	ρ=0.1	0.2	0.3	0.4	0.5	Average power
Cluster size 2						
PE	.099	.179	.295	.447	.620	-
s(.3)	.099	.179	.295	.447	.620	.3280
LM1	.099	.178	.295	.446	.618	.3074
LM2	.066	.112	.196	.324	.493	.2250
d^*	.082	.126	.187	.266	.365	.2052
D	.081	.125	.185	.262	.359	.1862
Cluster size 4						
PE	.140	.274	.433	.595	740	-
s(.3)	.140	.274	.433	.594	.738	.4358
LM1	.140	.272	.427	.582	.721	.4284
LM2	.095	.191	.326	.480	.632	.3448
d^*	.102	.178	.278	.398	.533	.2978
D	.101	.175	.274	.394	.528	.2944
Cluster size 8						
PE	.168	.309	.449	.579	.691	-
s(.3)	.167	.309	.449	.577	.690	.4384
LM1	.167	.303	.435	.555	.661	.4242
LM2	.121	.230	.352	.472	.585	.3520
d^*	.098	.167	.255	.357	.468	.2690
d	.097	.166	.254	.356	.466	.2678

Table 5.18. Powers of the Durbin-Watson, one- and two-sided Lagrange multiplier and POI-tests for the nx4 design matrix of cross-sectional data for 1961 and 1976, n=64, p=4 and α=.05

Tests	ρ=0.1	0.2	0.3	0.4	0.5	Average power
Cluster size 2						
PE	.134	.288	.505	.732	.900	-
s(.3)	.134	.288	.505	.732	.899	.5116
LM1	.134	.288	.505	.732	.899	.5116
LM2	.084	.191	.379	.618	.832	.4208
d^*	.102	.186	.307	.460	.630	.3370
D	.102	.185	.305	.456	.624	.3344
Cluster size 4						
PE	.220	.485	.731	.891	.966	-
s(.3)	.219	.485	.731	.890	.966	.6582
LM1	.220	.483	.728	.887	.963	.6562
LM2	.146	.373	.632	.830	.940	.5842
d^*	.143	.298	.493	.687	.842	.4926
D	.142	.295	.489	.683	.839	.4896
Cluster size 8						
PE	.297	.575	.772	.889	.952	-
s(.3)	.295	.574	.772	.889	.951	.6962
LM1	.296	.596	.763	.880	.944	.6958
LM2	.215	.473	.689	.833	.919	.6258
d^*	.148	.304	.486	.657	.797	.4784
D	.147	.303	.485	.657	.797	.4778

Table 5.19. Powers of the Durbin-Watson, one- and two-sided Lagrange multiplier and POI-tests for the nx4 design matrix of cross-sectional data for 1966 and 1971, n=64, p=4 and α=.05

Tests	ρ=0.1	0.2	0.3	0.4	0.5	Average power
Cluster size 2						
PE	.134	.288	.505	.732	.900	-
s(.3)	.134	.288	.505	.732	.899	.5116
LM1	.134	.288	.505	.731	.899	.5114
LM2	.084	.191	.379	.618	.832	.4208
d^*	.102	.186	.307	.460	.630	.3370
D	.102	.185	.305	.456	.624	.3344
Cluster size 4						
PE	.220	.486	.733	.891	.967	-
s(.3)	.220	.486	.733	.890	.966	.6592
LM1	.220	.484	.729	.888	.964	.6570
LM2	.147	.374	.634	.831	.940	.5852
d^*	.144	.299	.495	.689	.844	.4942
D	.143	.297	.492	.686	.841	.4918
Cluster size 8						
PE	.298	.577	.774	.890	.952	-
s(.3)	.296	.576	.774	.890	.952	.6976
LM1	.297	.571	.764	.881	.945	.6916
LM2	.216	.476	.691	.834	.920	.6274
d^*	.149	.306	.489	.661	.800	.4810
D	.149	.306	.489	.661	.801	.4812

Table 5.20. Powers of the Durbin-Watson, one- and two-sided Lagrange multiplier and POI-tests for the *n*x4 design matrix of cross-sectional data for 1961 and 1976, *n*=96, *p*=4 and *α*=.05

Tests	ρ=0.1	0.2	0.3	0.4	0.5	Average power
Cluster size 2						
PE	.165	.386	.664	.881	.977	-
s(.3)	.165	.386	.664	.881	.977	.6146
LM1	.165	.386	.664	.881	.977	.6146
LM2	.103	.272	.542	.806	.954	.5354
d*	.120	.242	.415	.616	.800	.4386
D	.120	.240	.412	.612	.796	.4360
Cluster size 4						
PE	.289	.639	.879	.973	.996	-
s(.3)	.288	.639	.879	.973	.996	.7550
LM1	.289	.638	.877	.972	.996	.7544
LM2	.198	.528	.814	.951	.992	.6966
d*	.178	.398	.648	.843	.950	.6034
D	.177	.396	.645	.840	.949	.6014
Cluster size 8						
PE	.406	.747	.913	.974	.993	-
s(.3)	.404	.747	.913	.974	.993	.8062
LM1	.404	.744	.909	.972	.992	.8042
LM2	.306	.660	.868	.956	.988	7556
d*	.194	.429	.664	.836	.935	.6116
d	.193	.427	.663	.835	.934	.6104

M. Ishaq Bhatti

Table 5.21. Powers of the Durbin-Watson, one- and two-sided Lagrange multiplier and POI-tests for the nx4 design matrix of cross-sectional data for 1966 and 1971, n=96, p=4 and α=.05

Tests	ρ=0.1	0.2	0.3	0.4	0.5	Average power
Cluster size 2						
PE	.165	.386	.664	.881	.977	-
s(.3)	.165	.386	.664	.881	.977	.6146
LM1	.165	.386	.664	.881	.977	.6146
LM2	.103	.272	.542	.806	.954	.5354
d^*	.120	.242	.415	.616	.800	.4386
D	.120	.241	.412	.612	.796	.4362
Cluster size 4						
PE	.289	.640	.879	.973	.996	-
s(.3)	.288	.640	.879	.973	.996	.7554
LM1	.289	.639	.877	.972	.996	.7546
LM2	.199	.529	.815	.951	.992	.6972
d^*	.178	.400	.650	.845	.951	.6048
D	.177	.398	.647	.843	.950	.6030
Cluster size 8						
PE	.408	.750	.914	.975	.994	-
s(.3)	.407	.750	.914	.975	.994	.8080
LM1	.407	.746	.911	.973	.993	.8060
LM2	.309	.664	.870	.957	.988	7576
d^*	.196	.434	.670	.840	.937	.6154
D	.196	.433	.669	.839	.936	.6146

6. TESTING FOR CLUSTER EFFECTS IN MSLR MODELS

6.1. INTRODUCTION [1]

The previous chapter of this book considered a 2SLR model, and compared the power of the POI test with that of the existing tests and the PE, for testing cluster effects within clusters. This chapter extends the 2SLR model to more general situations, by dividing each cluster further into subclusters and hence, it was named a 3SLR model in chapter two. When the procedure of further divisions of subclusters continues, it leads to a MSLR model. The correlated errors within clusters and subclusters give rise to intra-cluster correlation, ρ_1, and intra-subclusters correlation, ρ_2, respectively. The aim of this chapter is to derive some optimal tests for the twin problems of testing nonzero values of the intra-cluster, ρ_1, and of the intra-subcluster correlation, ρ_2, as follows:

1   1 $H_0: \rho_1 = \rho_2 = 0$, against $H_a: \rho_1 > 0, \rho_2 > 0$.
      and

2    $H_0: \rho_1 = 0, \rho_2 > 0$, against $H_a: \rho_1 > 0, \rho_2 > 0$.

For the first problem, the LMMPI and POI tests are derived, whereas for the second problem, the exact POI test and the asymptotic one-sided and

---

[1] Some of the findings reported in section 3.6 were published in Wu and Bhatti, (1994).

two-sided Lagrange multiplier (LM1 and LM2) tests are derived. The main objective of this section is to assess the accuracy of the asymptotic critical values, and to compare the powers of the asymptotic tests with those of the POI tests. This chapter extends upon the three-stage model to a multi-stage LR model, and deals with further LR complicated testing problems.

This chapter is structured as follows. In the ensuing section, the 3SLR model will be introduced, and the LMMPI and POI tests will be constructed for testing ρ_1 and ρ_2. The POI test is used to obtain the power envelope, which is then used as the benchmark in assessing the small sample power performance of the LMMPI test. An empirical power comparison of the LMMPI test with that of the POI test is also conducted, whilst in the following sections of this chapter, the problem of testing the null hypothesis of a zero intra-subcluster equicorrelation coefficient, ρ_2, in the presence of an intracluster equicorrelation coefficient, ρ_1, is considered. Following this, the testing procedure developed in the previous section is applied to selected design matrices to answer the following questions:

1   Do the POI tests when applied to small sample cases have superior sizes and power properties compared to those of asymptotic tests?
2   How well do the LM tests perform compared to the POI tests?

In the second last section of this chapter the 3SLR model is extended to the multi-stage model, and generalizes the optimal testing procedures for testing the multi-cluster effects, whilst the last section contains some concluding remarks.

6.2. THE 3SLR MODEL AND THE TESTS

The Model

This section extends the 2SLR model considered in chapter five to the situation where data is obtained from a three-stage cluster sampling design. Following the summary of the 3SLR model (2.5) of chapter two, it is supposed that the total of n observations are sampled from m first-stage clusters, with $m(i)$ second-stage subclusters from the ith cluster, and with $m(i,j)$ third stage observations from the jth subcluster of the ith cluster, such that

$n = \sum_{i=1}^{m} \sum_{j=1}^{m(i)} m(i,j)$. Thus, the 3SLR model can be expressed as

$$y_{ijk} = \sum_{\ell=1}^{p} \beta_\ell x_{ijk\ell} + u_{ijk} \qquad (6.1)$$

for observations $k = 1,2,...,m(i,j)$ from the jth subcluster, subcluster, $j = 1,2,...,m(i)$, from the ith cluster, $i = 1,2,...,m$, with y_{ijk} as the dependent variable, and p independent variables $x_{ijk\ell}$, $\ell = 1,2,...,p$, one of which may be a constant. The decomposition of the error term u_{ijk} can be written as

$$u_{ijk} = v_i + v_{ij} + v_{ijk}$$

where v_i is the ith cluster effect, v_{ij} is the jth subcluster effect in the ith cluster and v_{ijk} is the remaining random effect. These three components of u_{ijk} are assumed to be mutually independent and normally distributed with

$$E(v_i) = E(v_{ij}) = E(v_{ijk}) = 0 \qquad (6.2)$$

and

$$\text{var}(v_i) = \sigma_1^2, \ \text{var}(v_{ij}) = \sigma_2^2, \ \text{var}(v_{ijk}) = \sigma_3^2$$

so that

$$E(u_{ijk}) = 0$$

and

$$E\left(u_{ijk}u_{rst}\right) = \begin{cases} 0 & \text{for } i^1r \qquad \text{any } j,s,k,t \\ \sigma_1^2 & \text{for } i = r, \ j^1s, \ \text{any } k,t \\ \sigma_1^2 + \sigma_2^2 & \text{for } i = r, \ j = s, \ k^1t \\ \sigma_1^2 + \sigma_2^2 + \sigma_3^2 & \text{for } i = r, \ j = s, \ k = t \end{cases} \qquad (6.3)$$

for $k,t = 1,2,...,m(i,j)$,

$j,s = 1,2,...,m(i)$

$i,r = 1,2,...,m$.

This gives rise to intra-cluster correlation, whose coefficient is

$$\rho_1 = \frac{\sigma_1^2}{\sigma^2} \qquad (6.4)$$

and intra-subcluster correlation, with the coefficient

$$\rho_2 = \frac{\sigma_1^2}{\sigma^2} \qquad (6.5)$$

where $\sigma^2 = \sigma_1^2 + \sigma_2^2 + \sigma_3^2$, so $0 \le \rho_1, \rho_2 \le 1, \rho_1 + \rho_2 \le 1$. Obviously, ρ_1 and ρ_2 originate from v_i and v_{ij} in error term u_{ijk}.

The regression model (6.1), under (6.2), (6.3), (6.4) and (6.5), can be written more compactly in matrix form as[2]

$$y = X\beta + u \qquad (6.6)$$

in which y and u are $n \times 1$, X is $n \times p$, β is $p \times 1$ and

$$u \sim N\left(0, \sigma^2\Omega\left(\rho_1, \rho_2\right)\right) \qquad (6.7)$$

---

[2] u is formed as $u = (u_{111}, \ u_{112}...u_{mm(i)m(i,j)})'$ where X and y can also be formed in a similar pattern.

where $\Omega(\rho_1,\rho_2) = \bigoplus_{i=1}^{m}\Omega_i(\rho_1,\rho_2)$ is cluster diagonal, with submatrices defined by (3.26) or by (3.27).

Although calculations can be undertaken with unbalanced data,[3] for the convenience of this discussion, and in relation with chapters three and four, the remainder of this chapter is constrained to situations where data is balanced over clusters and subclusters. So let $T = m(i,j)$, and further assume $s = m(i)$, $i = 1,...,m$, $j = 1,...,s$, then $\Omega(\rho_1, \rho_2)$ in (6.7) can be simplified in terms of the usual notation of D_1 and D_2 as follows,

$$\Omega(\rho_1,\rho_2) = (1 - \rho_1 - \rho_2)I_n + \rho_1 D_1 + \rho_2 D_2 \qquad (6.8)$$

for which D_1 and D_2 are as in (3.31) and I_n is $n \times n$ identity matrix, such that $n = msT$, and where, m is the number of main clusters, s is the number of subclusters in each cluster and T stands for the number of observations in each subcluster of the main cluster.

6.3. THE TESTS

Locally Most Mean Powerful Invariant Test

In this section derives the LMMI and POI tests for testing (6.9) against (6.10) defined below. It is important to note that the error covariance matrix in (6.7) involves two unknown parameters, ρ_1 and ρ_2. The main problem of interest of this subsection is to test

$$H_0 : \rho_1 = \rho_2 = 0 \qquad (6.9)$$

against

$$H_a : \rho_1 \geq 0, \rho_2 \geq 0, \text{ (excluding } H_0). \qquad (6.10)$$

---

[3] That is, different number of observations in each subcluster and different number of subcluster in each main cluster.

This testing problem is invariant, with respect to transformations of the form

$$y \rightarrow r_0 y + Xr, \tag{6.11}$$

where r_0 is a positive scalar and r is a $p \times 1$ vector. The $m \times 1$ vector

$$\vartheta = \frac{Py}{\left(y'P'Py'\right)^{1/2}} \tag{6.12}$$

is a maximal invariant under this group of transformation, where $m' = n-p$, $M_x = I_n - X(X'X)^{-1}X'$, and P is an $m' \times n$ matrix such that $PP' = I_{m'}$ and $P'P = M_x$. The probability density function of ϑ under (6.7) and (6.8) can be shown as

$$f\left(\vartheta, \rho_1, \rho_2\right) = 1/2\Gamma\left(\frac{m'}{2}\right)\pi^{-m'/2}\left|P\Omega\left(\rho_1, \rho_2\right)P'\right|^{-1/2}$$
$$\left(\vartheta'\left(P\Omega\left(\rho_1, \rho_2\right)P'\right)^{-1}\vartheta\right)^{-m'/2} d\vartheta, \tag{6.13}$$

where $d\vartheta$ denotes the uniform measure on the unit sphere. Note that the only unknown parameters in (6.13) are ρ_1 and ρ_2.

Based on ϑ, when more than one parameter is being tested, King and Wu (1990) showed that a locally most mean powerful test for

$$H_0' : \theta = 0$$

against

$$H_a' : \theta_1 \geq 0, ..., \theta_p \geq 0_p, \theta \neq 0,$$

is to reject H_0' for small values of

$$d = \frac{\hat{u}'A\hat{u}}{\hat{u}'\hat{u}}, \tag{6.14}$$

where θ is $p \times 1$ vector of parameters,

$$A = \sum_{i=1}^{p} A_i,$$

$$A_i = -\frac{\partial \Sigma(\theta)}{\partial \theta_i}\bigg|_{\theta=0}, \quad i = 1, \ldots, p, \tag{6.15}$$

and $\hat{u}$ is a vector of OLS residuals from (6.7). When $p=1$, d in (6.14) reduces to the locally best invariant (LBI) test statistic proposed by King and Hillier (1985). For the problem of testing (6.9) against (6.10), the LMMPI test is determined by (6.14), with $A = A_1 + A_2$ is where

$$A_1 = \bigoplus_{i=1}^{m} A_{1i} = \left(I_n - D_1 \right)$$

$$A_2 = \bigoplus_{i=1}^{m} A_{2i} = \left(I_n - D_2 \right)$$

and as noted earlier in (3.31)

$$A = \left(2I_n - D_1 - D_2 \right).$$

Therefore d in (6.14) can be expressed and simplified as

$$d = \frac{\hat{u}' A \hat{u}}{\hat{u}' \hat{u}} = \frac{\hat{u}' \left(2I_n - D_1 - D_2 \right) \hat{u}}{\hat{u}' \hat{u}}$$

$$= 2 - \left[\left(\hat{u}' D_1 \hat{u} + \hat{u}' D_2 \hat{u} \right) / \hat{u}' \hat{u} \right]$$

$$= 2 - \sum_{i=1}^{m} \left(\sum_{j=1}^{s} \sum_{k=1}^{T} \hat{u}_{ijk} \right)^2 / \sum_{i=1}^{m} \sum_{j=1}^{s} \sum_{k=1}^{T} \hat{u}_{ijk}^2$$

$$-\sum_{i=1}^{m}\sum_{j=1}^{s}\left(\sum_{k=1}^{T}\hat{u}_{ijk}\right)^{2} / \sum_{i=1}^{m}\sum_{j=1}^{s}\sum_{k=1}^{T}\hat{u}_{ijk}^{2} \, . \tag{6.16}$$

Note expression of (6.16) is a ratio of quadratic forms in normal variates, so its critical values can be evaluated by using the standard numerical techniques that were mentioned in earlier chapters. In addition to this if $T = 0$, then d in (6.16) becomes d' in a form given by (5.15), which implies that the data does not possess the subcluster effects, and hence one would prefer to use d' test of (5.15) given in chapter five.

Point Optimal Invariant Test

For the testing problem in the previous subsection, King's (1987b) POI test can be used to obtain the maximum attainable power, i.e. the power envelope over the alternative hypothesis parameter space. For any particular test, its performance can be assessed by checking how close its power is to the PE. The POI test statistic is introduced in this section, and used in the following subsection in the comparative power study assessing the power performance of the LMMPI test, which was derived in the previous subsection. Also note that this testing problem is invariant under the group of transformation (6.11), with the maximal invariant vector, ϑ, given by (6.12) and the density of ϑ, by (6.13).

Suppose $\left(\rho_{1},\rho_{2}\right)' = \left(\rho_{1}^{*},\rho_{2}^{*}\right)'$ is a chosen point in the alternative parameter space. Based on the maximal invariant in (6.12), one can obtain the POI test for testing (6.9) against (6.10), that is, to reject the null hypothesis for small values of

$$s = \frac{\hat{u}'\Delta\hat{u}}{\hat{u}'\hat{u}} \tag{6.17}$$

where

$$\Delta = \Omega^{*-1} - \Omega^{*-1} X \left(X'\Omega^{*-1} X\right)^{-1} X'\Omega^{*-1},$$

$$\Omega^* = \Omega\left(\rho_1^*, \rho_2^*\right)$$

and $\hat{u}$ again is the vector of OLS residuals, i.e. $\hat{u} = M_x y$. Thus, tests based on s give the maximum attainable power at $\left(\rho_1^*, \rho_2^*\right)$ in the class of invariant tests.

Note that s in (6.17) is a ratio of quadratic forms in normal variates, so its critical values can be evaluated using GUASS subroutines stated earlier. To calculate power, the error covariance matrix, $\Omega(\rho_1, \rho_2)$ in (6.8), or its diagonal component matrix $\Omega_i(\rho_1, \rho_2)$ in (3.27), needs to be decomposed such that

$$\Omega_i\left(\rho_1, \rho_2\right) = \Omega_i^{1/2}\left(\rho_1, \rho_2\right)\left(\Omega_i^{1/2}\left(\rho_1, \rho_2\right)\right)',$$

where the formula for $\Omega_i^{1/2}\left(\rho_1, \rho_2\right)$ is given by (3.35).

It can be observed that the power of the critical region $s < c_\alpha$ in (6.17) can be found by evaluating probabilities of the form

$$\Pr\left[s < c_\alpha\right] = \Pr\left[u'\left(\Delta - c_\alpha M_x\right)u < 0 \middle| u \sim N\left(0, \sigma^2 \Omega\left(\rho_1, \rho_2\right)\right)\right]$$

$$= \Pr\left[\left(\Omega^{-1/2}\left(\rho_1, \rho_2\right)u\right)'\left(\Omega^{1/2}\left(\rho_1, \rho_2\right)\right)'\left(\Delta - c_\alpha M_x\right)\Omega^{1/2}\right.$$

$$\left.\left(\rho_1, \rho_2\right)\left(\Omega^{-1/2}\left(\rho_1, \rho_2\right)u\right) < 0 \middle| \Omega^{-1/2}\left(\rho_1, \rho_2\right)u \sim N\left(0, \sigma^2 I_n\right)\right]$$

$$= \Pr\left[\sum_{i=1}^{n} \lambda_i \xi_i^2 < 0\right], \tag{6.18}$$

where $\lambda_1, \lambda_2, \ldots, \lambda_n$ are the eigenvalues of

$$\left(\Omega^{1/2}\left(\rho_1, \rho_2\right)\right)'\left(\Delta - c_\alpha M_X\right)\left(\Omega^{1/2}\left(\rho_1, \rho_2\right)\right), \tag{6.19}$$

and $\xi_1^2, ..., \xi_n^2$ are independent chi-squared random variables, each with one degree of freedom. In (6.19), the matrix is given by

$$\Omega^{1/2}\left(\rho_1, \rho_2\right) = \overset{m}{\underset{i=1}{\oplus}} \Omega_i^{1/2}\left(\rho_1, \rho_2\right),$$

where the elements of the matrix, $\Omega_i^{1/2}\left(\rho_1, \rho_2\right)$ is formed by the formula (3.35).

6.4. POWER COMPARISONS OF THE POI AND THE LMMPI TESTS

This section reports a summary of the results of the Wu and Bhatti's (1994) empirical power comparison on the performance of the LMMPI test for intra-cluster and intra-subcluster equicorrelation coefficients. The POI test discussed in the previous section will be used here to get the maximum obtainable power at selected points in the alternative parameter space. The differences between this power envelope and the power of the LMMPI test will determine the usefulness of the latter test.

Two sets of design matrices are used in this study:

- $X1$ (p=3): Bangladesh agricultural survey data, including a constant, log of labour input per acre in man-days, and log of biological-chemical (BC) input per acre in thousand of Bangladesh Taka.[4]
- $X2$ (p=4): Artificially generated data, consisting of a constant, two independent log-normal variables and one uniform random variable. The values of these variables are presented in table 6.3 of appendix 6A.

Here, $X1$ represents some typical economic survey data whose details are given in tables 6.1 and 6.2 of appendix 6A. For this data, two cluster structures are considered, with the first structure involving two first-stage clusters, with each of them having two subclusters respectively, so that $N_i = 2$ for $i = 1,2$. The second structure whereas also involves two first-stage clusters, with each of

---

[4] Thanks are due to Dr. Asraul Hoque for making this data available.

them having three subclusters, so that $N_i= 3$ for $I =1,2$. Also, two different sample sizes, $n = 48$, $n = 96$, are considered, so that with two cluster structures, and two sample sizes, four cases are associated with $X1$, $X2$:

    a. $N_i xT = 2x12,$ $i = 1,2,$ $n = 48$

    b. $N_i xT = 3x8,$ $i = 1,2,$ $n = 48$

    c. $N_i xT = 2x24,$ $i = 1,2,$ $n = 96$

    d. $N_i xT = 3x16,$ $i = 1,2,$ $n = 96$.

The $X1$ data follow the above experimental design, whereas the $X2$ data is selected from a set of 96 observations in all cases.

The powers are evaluated at points of combinations of $\rho_1 = 0.0$, 0.05, 0.1, 0.2, 0.4 and $\rho_2 = 0.0$, 0.05, 0.1, 0.2, 0.4. Since ρ_1 and ρ_2 obey the constraint $\rho_1 + \rho_2 \le 1$, and weak to medium equicorrelation is likely, as noted by Scott and Holt (1982) in the two-stage model context, the maximum values of ρ_1 and ρ_2 have been set at 0.4. For each case, the power envelope is evaluated at three points: $(\rho_1, \rho_2)' = (0.0, 0.2)'$ $(0.2, 0.0)'$ and $(0.2, 0.2)'$. The critical values and the powers are calculated using a modified version of Koerts and Abrahamse's (1969) FQUAD subroutine. Table 6.4 presents the results for $X1a$ - $X1d$, $X2a$ and $X2b$. Since power patterns in the cases of $X2c$ and $X2d$ are quite similar to those of $X1c$ and $X1d$, their results are omitted.

From these tables, it is obvious that the LMMPI test has quite good power for the situations considered. Out of 18 points (three for each case) in the alternative space, where the power envelope is calculated, the LMMPI test had power differing from the envelope by less than 0.12 at 17 points, to differing by less than 0.05 at 10 points. Since the LMMPI test has local average optimal power, these calculated powers may be considered reasonable results and are encouraging. Comparing case $X1a$ to $X1c$ and case $X1b$ to case $X1d$, where sample size has been doubled, the power has increased significantly. Also, with this increased sample size, the differences between the power of the LMMPI test, and the power envelope have decreased. Given that these differences are small in the first instance, this decrease is relatively significant. In all cases where the number of subcluster is raised from $N_i=2$ to $N_i=3$ while the total number of observations is unchanged, i.e., from case $X1a$ to $X1b$, $X1c$ to $X1d$ and $X2a$ to $X2b$, and the power has dropped for combinations of small to moderately small values of ρ_1 and ρ_2. However, this pattern is reversed for combinations of ρ_1 and ρ_2, with $\rho_2 = 0.4$ for cases involving $X1$ data. For $X2$ data, this reverse happened at $(\rho_1, \rho_2)' = (0.4, 0.4)'$, with the furthest point from the null hypothesis considered in our study.

6.5. TESTING ρ_2 IN THE PRESENCE OF ρ_1 [5]

The main purpose of this section is to consider the problem of testing subcluster effects in the presence of main cluster effects, by applying King's (1989) approach to the 3SLR model with equicorrelated disturbances in clusters and within subclusters. Such testing procedures have many applications in physical and social sciences. For example in socioeconomic surveys, a researcher may be interested in testing for the presence of random errors of key punch operators in the presence of coding, or for an interviewer's error. In time series cross-sectional data econometricians may be interested in testing firm effects in the presence of industry effects, and so on. Note that this testing problem is similar to that of King (1989) who investigated the problem of testing for an AR(4) regression disturbance process in the presence of an AR(1) process. In this study the objective is to construct POI, LM1 and LM2 tests to detect the subcluster effects in the presence of main cluster effects.

6.6. THE 3SLR MODEL AND THE NEW TESTS

This section considers the 3SLR model discussed in (6.6) in the previous sections, i.e.,

$$y = X\beta + u, \ u \sim N\left(0, \sigma^2 \Omega(\rho_1, \rho_2)\right),$$

where

---

[5] Some of the findings reported in this section were presented at the 1992 AustralasianMeeting of the Econometrics Society, Department of Econometrics, Monash University, Clayton, Australia, July 6-8, 1992. The title of the paper was 'Testing forsubblock effect in the multistage linear regression models'.

$$\Omega\left(\rho_1,\rho_2\right)=\bigoplus_{i=1}^{m}\Omega_i\left(\rho_1,\rho_2\right),$$

with submatrices given by (3.27), in which $0\leq\rho_1+\rho_2\leq1$ and ρ_1, ρ_2 are unknown parameters. For the purpose of test construction, there are a couple of options to decide on the range of ρ_1 values. The first option is that $0\leq\rho_1\leq1$, which seems unnecessarily generous, whilst the other is possibly more reasonable, in that to assume that $0\leq\rho_1\leq0.5$, because ρ_1 and ρ_2 follow the constraint $0\leq\rho_1+\rho_2\leq1$.

The aim in this section is to develop and discuss testing the complicated null hypothesis of the form,

$$H_0:\rho_2=0,\rho_1>0 \tag{6.20}$$

against

$$H_a:\rho_2>0,\rho_1>0. \tag{6.21}$$

Using invariance arguments, the nuisance parameters except for ρ_1 can be eliminated, and thus the testing problem (6.20) against (6.21) simplifies to one of testing

$$H_0:\vartheta \text{ has the density } f\left(\vartheta,\rho_1,0\right),0<\rho_1\leq0.5$$

against

$$H_a:\vartheta \text{ has the density } f\left(\vartheta,\rho_1,\rho_2\right),0\leq\rho_1\leq0.5,\ 0\leq\rho_2\leq0.5.$$

Point optimal invariant test
First consider the simple problem of testing the hypothesis

$$H_0':\left(\rho_1,\rho_2\right)'=\left(\rho_{10},0\right)'$$

against the simple alternative hypothesis

$$H'_a : (\rho_1, \rho_2)' = (\rho_{11}, \rho_{21})',$$

where $0 \le \rho_{10} \le 0.5, 0 \le \rho_{11} \le 0.5 \; and \; 0 \le \rho_{21} \le 0.5$ are known and fixed.
Suppose under $H_0, \Omega_0 = \Omega(\rho_{10}, 0)$ and under $H_a, \Omega_1 = \Omega(\rho_{11}, \rho_{21})$.

The Neyman-Pearson lemma implies that a Most Powerful (MP) test within the class of invariant tests can be based on critical regions of the form

$$s(\rho_{10}, \rho_{11}, \rho_{21}) = \frac{\vartheta'(P\Omega_1 P')^{-1} \vartheta}{\vartheta'(P\Omega_0 P')^{-1} \vartheta} < c_\alpha$$

where c_α is an appropriate critical value. King (1989, p. 290) showed that the statistics $s(\rho_{10}, \rho_{11}, \rho_{21})$ can also be written as

$$s(\rho_{10}, \rho_{11}, \rho_{21}) = \frac{u'\Delta_1 u}{u'\Delta_0 u} \tag{6.22}$$

where

$$\Delta_1 = \Omega_1^{-1} - \Omega_1^{-1} X (X'\Omega_1^{-1} X)^{-1} X'\Omega_1^{-1}$$

$$\Delta_0 = \Omega_0^{-1} - \Omega_0^{-1} X (X'\Omega_0^{-1} X)^{-1} X'\Omega_0^{-1}.$$

Note that $s(\rho_{10}, \rho_{11}, \rho_{21})$ in (6.22) is a ratio of quadratic forms in normal variables, and therefore its critical values, c_α, for a desired level of significance, α, may be obtained by evaluating the probabilities of the form

$$\Pr\left[s(\rho_{10}, \rho_{11}, \rho_{21}) < c_\alpha \,\big|\, u \sim N(0, \Omega_0) \right]$$

$$= \Pr\left[u'(\Delta_1 - c_\alpha \Delta_0) u < 0 \,\big|\, u \sim N(0, \Omega_0) \right]$$

$$= \Pr\left[\left(\Omega_0^{-1/2}u\right)'\left(\Omega_0^{1/2}\right)'\left(\Delta_1 - c_\alpha\Delta_0\right)\Omega_0^{1/2}\left(\Omega_0^{1/2}u\right) < 0 \Big| u \sim N\left(0,\Delta_0\right)\right]$$

$$= \Pr\left[\sum_{i=1}^{n}\lambda_i\xi_i^2 < 0\right], \tag{6.23}$$

where $\lambda_1,...,\lambda_n$ are eigenvalues of the matrix

$$\left(\Omega_0^{1/2}\right)'\left(\Delta_1 - c_\alpha\Delta_0\right)\left(\Delta_0^{1/2}\right),$$

and $\xi_1^2,...,\xi_n^2$ are independent chi-squared random variables, each with one degree of freedom.

Thus a test based on $s\left(\rho_0,\rho_1,\rho_2\right)$ in (6.22) is most powerful invariant in the neighbourhood of $\left(\rho_1,\rho_2\right)' = \left(\rho_{11},\rho_{21}\right)'$. Therefore it is a POI test of the simple null H_0' against the simple alternative H_a'. For the wider problem of testing (6.20) against (6.21) it may not be necessarily by POI, as for this testing problem the critical value should be found by solving

$$\mathop{Sup}_{0<\rho_1\leq0.5} \Pr\left[s\left(\rho_{10},\rho_{11},\rho_{21}\right) < c_\alpha^* \Big| u \sim N\left(0,\Omega\left(\rho_1,0\right)\right)\right] = \alpha \ for \ c_\alpha^*.$$

In general, $c_\alpha < c_\alpha^*$ so the two critical regions are different. As the range of ρ_1 is closed

$$\Pr\left[s\left(\rho_{10},\rho_{11},\rho_{21}\right) < c_\alpha^* \Big| u \sim N\left(0,\Omega\left(\rho_1,0\right)\right)\right] = \alpha$$

is true for at least one ρ_1 value. If we can find such a ρ_{10} value to let $c_\alpha = c_\alpha^*$ then the test based on $s\left(\rho_{10},\rho_{11},\rho_{21}\right)$ is MPI, and is a POI test of (6.20) against (6.21).

However, in general, there is no reason to believe that such a ρ_{10} value should exist, and though perhaps it may exist for some combinations of X

matrices and $\left(\rho_{11},\rho_{21}\right)'$ values, it may not for others. For the X matrices used in this experiment such a ρ_{10} does exist, and hence the POI test also exists for testing (6.20) against (6.21).

Once the value of $\rho_{10} = \rho_{10}^*$ (say) is chosen, the optimal values of ρ11, ρ21 can be decided upon in the same way as in King (1989). This implies that for any fixed ρ21, the power of the test will be determined in part by the choice of ρ11. This study suggests choosing ρ_{11} to maximize the minimum power in the alternative parameter space i.e.

$$\max_{0 \le \rho_1 \le 0.5}\left[\min \Pr\left\{s\left(\rho_{10}^*,\rho_{11},\rho_{21}\right) < c_\alpha^* \middle| u \sim N\left(0,\Omega\left(\rho_{11},\rho_{21}\right)\right)\right\}\right].$$

This ρ_{11} value is denoted by ρ_{11}^*. Then the optimal value of ρ_{21} can be found by solving

$$\Pr\left[s\left(\rho_{10}^*,\rho_{11}^*,\rho_{21}\right) < c* \middle| u \sim N\left(0,\Omega\left(\rho_{11}^*,\rho_{21}\right)\right)\right] = p, \tag{6.24}$$

for $\rho_{21} = \rho_{21}^*$, where p on the right hand side of (6.24) denotes the requisite level of the power. It may be noted that a simpler form of this exercise in the context of a BO test has already been undertaken in chapter three. This is another example of applying the same logic to a more complex set of given parameters, under the alternative parameter space.

In the next section of this chapter the procedure of calculating critical values and the power of the POI test for some selected design matrices will be discussed. This is done in order to compare the power performance of a POI test with its other competitive tests, namely LM1 and LM2 tests, as it has been done in chapter five. In the next subsection the derivation of these tests is demonstrated, and the power comparison is shown in a later section of this chapter.

Lagrange Multiplier Test

The Lagrange multiplier test is generally a two-sided test, and is only valid asymptotically. In other words it relaxes the constraints on the sign of the parameters under the alternative hypothesis. The main advantage of the LM test over its asymptotic equivalent tests, namely, the LR and Wald tests, is that it requires an estimate of the value of the parameters only under the null hypothesis. Due to this factor, it has gained wide acceptance from practitioners in the econometrics literature. Examples of its use include Breusch and Pagan (1980), Honda (1985), Godfrey (1988), Baltagi, Chang and Li (1992) and Wu (1991). For a survey of general econometric methods, see Pagan and Wickens (1989). Krämer and Sonnberger (1986) provide some detailed descriptions of various tests in the context of the linear regression model. In this section the LM1 and LM2 tests will be considered, because as was noted they require the estimation of ρ_1 under the null hypothesis only.

Corresponding to the testing problem of (6.20) against (6.21), consider the problem of testing

$$H_0' : \rho_2 = 0, \rho_1 \neq 0$$

against

$$H_a' : \rho_2 \neq 0, \rho_1 \neq 0.$$

The log likelihood function for the model (6.6) is

$$L\left(\beta, \sigma^2, \rho_1, \rho_2\right) = -\frac{n}{2}\ell n\left(2\pi\sigma^2\right) - \frac{1}{2}\ell n\left|\Omega\left(\rho_1, \rho_2\right)\right|$$

$$-\frac{1}{2\sigma^2}\left(y - X\beta\right)'\Omega^{-1}\left(\rho_1, \rho_2\right)\left(y - X\beta\right). \tag{6.25}$$

It can be verified that

$$\frac{\partial L}{\partial \rho_2}\bigg|_{H_0} = d_1 - \frac{n}{2}\frac{\hat{u}'A\hat{u}}{\hat{u}'\hat{u}} \tag{6.26}$$

in (6.26) which $d_1 = -\frac{1}{2}\frac{\partial\left|\Omega(\rho_1,\rho_2)\right|}{\partial\rho_2}\cdot\left|\Omega(\rho_1,\rho_2)\right|^{-1}\Big|_{H_0}$

or equivalently,

$$\frac{n(T-1)b}{2},$$

where

$$b = \frac{-\rho}{(1-\rho_1)(1-\rho_1+sT\rho_1)},$$

and A in (6.26) is expressed as,

$$A = \Omega^{-1/2}(\rho_1,0)(I_n - D_2)\left(\Omega^{-1/2}(\rho_1,0)\right)'.$$

The information matrix ψ_θ of the linear regression model (6.6) can be obtained by using Magnus (1978, theorem 3), where ψ_θ is a symmetric 4x4 matrix with typical element

$$(\psi_\theta)_{ij} = \frac{1}{2}tr\left(\frac{\partial\Omega^{-1}}{\partial\theta_i}\Omega\frac{\partial\Omega^{-1}}{\partial\theta_j}\Omega\right).$$

Let

$$\Omega(\theta) = \sigma^2\Omega(\rho_1,\rho_2),$$

where

$$\theta = (\theta_1,\theta_2,\theta_3)' = (\sigma^2,\rho_1,\rho_2)'.$$

Thus the inverse of our cluster diagonal information matrix evaluated at the null is

$$
\psi^{-1}\big|H_0 = \begin{bmatrix} \left(X'\Omega^{-1}X\right)^{-1} & 0 & 0 & 0 \\ & \begin{bmatrix} \psi_{11} & \psi_{12} & \psi_{13} \\ \psi_{21} & \psi_{22} & \psi_{23} \\ \psi_{31} & \psi_{32} & \psi_{33} \end{bmatrix}^{-1} \\ 0 & & & \\ \begin{matrix}0\\0\end{matrix} & & & \end{bmatrix}_{\big|H_0} .
$$

Let I_0 be denoted as the bottom right element of ψ^{-1} which corresponds to ρ_2, i.e.

$$
I_0 = \left[\psi_{33} - \left(\psi_{31}\ \psi_{32}\right) \begin{bmatrix} \psi_{11} & \psi_{12} \\ \psi_{12} & \psi_{22} \end{bmatrix}^{-1} \begin{pmatrix} \psi_{13} \\ \psi_{23} \end{pmatrix} \right]^{-1}_{\big|H_0} ,
$$

which can be simplified as

$$
I_0 = \left[\psi_{33} - \frac{1}{\left(\psi_{11}\psi_{22} - \psi_{12}^2\right)} \left\{ \psi_{13}^2\psi_{22} - 2\psi_{12}\psi_{13}\psi_{23} + \psi_{11}\psi_{23}^2 \right\} \right]^{-1}_{\big|H_0} \qquad .(6.27)
$$

Hence the I_0 in (6.27) is obtained by noting that

$$
\psi_{ij} = \psi_{ji}, \ i,j = 1,2,3, \ for \ i \neq j, i < j.
$$

Also note that the elements of ψ_{ij} under the null hypothesis are defined as follow,

$$
\psi_{11} = \frac{n}{2\sigma^4}, \psi_{12} = \frac{nb(sT-1)}{2\sigma^2}, \psi_{13} = \frac{nb(T-1)}{2\sigma^2}
$$

$$\psi_{22} = \frac{n(1-sT)}{2}\left[-a^2 + b^2sT(1-sT) + 2ab(1-sT)\right],$$

$$\psi_{32} = \frac{n(1-T)}{2}\left[-a^2 + b^2sT(1-sT) + 2ab(1-sT)\right],$$

$$\psi_{33} = \frac{n(1-T)}{2}\left[-a^2 + b^2sT(1-T) + 2ab(1-T)\right], \qquad (6.28)$$

where

$$a = \frac{1}{1-\rho_1} \text{ and } b \text{ is defined previously by } (6.26).$$

Thus, by (6.26), (6.27) and (6.28), the LM1 test statistic is

$$s1 = \frac{\partial L}{\partial \rho_2}\bigg|_{H_0}\bigg/\sqrt{I_0} , \qquad (6.29)$$

and the LM2 test statistic is

$$s2 = \left(\frac{\partial L}{\partial \rho_2}\bigg|_{H_0}\right)^2\bigg/I_0 . \qquad (6.30)$$

The asymptotic distributions of the $s1$ and $s2$ test statistics under H_0 are standard normal and chi-square, with one degree of freedom respectively. Note that under H_0, the ρ_1, σ^2 and β are replaced by their respective maximum likelihood estimates, namely $\hat{\rho}_1, \hat{\sigma}^2$ and $\hat{\beta}$. Therefore, the ψ_{ij}'s will also be replaced by $\hat{\psi}_{ij}$'s for all values of i and j defined above. The next section of this chapter will report the results of the Monte Carlo study designed to compare the empirical power of the one-sided and two-sided Lagrange multiplier tests with that of the two versions of the POI test for testing intra-subcluster equicorrelation in the presence of intracluster equicorrelation for the various selected design matrices.

6.7. POWERS AND SIZES COMPARISON OF LM1, LM2 AND POI TESTS

This section reports the results of Monte Carlo experiments which is conducted to assess and compare the small-sample size and power performance of the two versions of the POI tests, namely $s\left(\rho_{10}^*, 0.1, 0.1\right), s\left(\rho_{10}^*, 0.25, 0.25\right)$ and the LM1 and the LM2 tests of testing null hypothesis of (6.20) against an alternative (6.21). The main objective of this analysis is to assess the validity (or accuracy) of the asymptotic critical values for each test in finite samples, and to compare the power curves of the asymptotic tests (i.e. LM1 and LM2) with the two versions of the POI tests.

Experimental Design and the Computations

This experiment is divided into two parts. The first part is used the Monte Carlo method to estimate probabilities of a type I error under H_0 at $\rho = 0.0$, 0.05, 0.1, 0.2, 0.3, 0.4 and 0.5, for the LM1 and LM2 tests at the five percent nominal level. The second part of the experiment whereas is completed in two stages, with the first stage being the calculation of appropriate critical values so that the comparison of powers, which constitute the second stage, can be made at approximately the same level of significance. Critical values of ρ_{10}^* for the POI tests were calculated according to the procedure given previously for all X matrices defined below, with their calculated values are tabulated in table 6.5 of appendix 6B. For the LM1 and LM2 tests, the Monte Carlo method was used to estimate exact-size critical values at $\rho_1 = 0.0, 0.05, 0.1$, 0.2, 0.3, 0.4 and 0.5 under H_0 at the five percent level. From each set of seven critical values, the largest was selected, thus ensuring that at least at the chosen points, the size of the test does not exceed five percent. The small-sample sizes and powers were then calculated using the methodology discussed for the POI tests, with the methodology for the LM1 and LM2 tests given earlier in the subsection entitled Lagrange Multiplier Test. Sizes were calculated for the equicorrelated parameter, ρ_1, values 0.0, 0.05,..., 0.5 and powers were calculated at $\rho_1 = \rho_2 = 0.0, 0.05,...,0.5$. For the small-samples, the sizes and the powers of the $s(\rho_{10}^*, 0.1, 0.1)$ and $s(\rho_{10}^*, 0.25, 0.25)$ tests were calculated at ρ_1 and ρ_2 values of 0.0, 0.05, 0.1, 0.2, 0.3, 0.4 and 0.5. In the remainder of this section these tests will be denoted as the $s(0.1, 0.1)$ and $s(0.25, 0.25)$ tests.

The following design matrices were used in this study:

Bangladesh agricultural survey data,[6]

$X1$: (nx3; n=24, m=2, s=3, T=4)

(nx3, n=48, m=2, s=3, T=8)

Australian data[7]

$X2$: (nx3; n=24, m=2, s=3, T=4)

(nx3; n=64, m=2, s=4, T=8)

Artificially generated data

$X3$: (nx3; n=24, m=2, s=3, T=4)

(nx3; n=72, m=2, s=3, T=12).

These design matrices reflect a variety of economic phenomenon, and have also been used in earlier empirical studies (see Hoque: 1988, 1991; King and Evans: 1986; Wu and Bhatti: 1994; Wu: 1991 and Bhatti (2004) among others).

The Monte Carlo method is applied using two thousand replications for the 1m tests. The results of Breusch (1980) imply that the sizes and powers of each of the tests are invariant to the values taken by β and σ^2, and therefore, for the application of the Monte Carlo method, β_i, i =1,2,...,p and σ^2 were all set to unity. Whenever needed, the innovations, v_t, are generated as pseudo-

---

[6] The values of the variables, details of the sampling design and measurement of variables are given in appendix 6A. This design matrix has also been used in the empirical study of the section entitled 'Power comparison of the POI and LMMPltests' with sample of size 96.
[7] The detailed explanation of this design matrix is already given in chapter five.

random normal variables as described by King and Giles (1984). These innovations are then transformed to the equicorrelated error structure of the form u_{ijk}, defined by (6.1) using a subroutine under H_0, following the logic of Beach and MacKinnon's (1978), but otherwise different from them.[8] The maximum likelihood estimates of β, σ^2, and ρ_1 under H_0 were computed using Ansley's (1979) approach for estimating ARMA process, applied to the 2SLR model. For an analogous application of Ansley's algorithm to the linear regression model and the 2SLR model with varying coefficients, see King (1986) and Bhatti (1993). This application reduces the maximization problem to a sum of squares minimization problem, which is handled by the IMSL subroutine DBCLSF from the IMSL MATH/LIBRARY (1989), with the constraint that $|\rho_1| \le 1$. For the POI test, the probabilities of the form (6.23) and (6.24) are calculated using the eigenvalue method and the modified version of Koerts and Abrahamse's (1969) FQUAD subroutine, with maximum integration and truncation errors of 10^{-6}.

The Results

Tables 6.6 and 6.7 in appendix 6B report the estimated sizes for the various sample-sizes over ρ_1= 0.0, 0.05, 0.1, 0.2, 0.3, 0.4, 0.5 of the LM1 and LM2 test statistics, using asymptotic critical values at the five percent significance level. In these tables, it can be noticed that almost all of the estimated sizes of the LM1 and LM2 tests for the different sample sizes are significantly different than the nominal sizes of 0.05 for the X1, X2 and X3 design matrices. The exceptions to this are in table 6.6, for the design matrices X1 for ρ_1= 0.05, 0.1, 0.2, X2 for ρ_1= 0.0, 0.5 and X3 for ρ_1= 0.0, 0.05 and 0.1, where the estimated sizes of the LM2 test did not differ significantly from the nominal size. In all cases, the estimated sizes of the LM2 test are more than half of the nominal level, except in table 6.7 where for the design matrix X2 for ρ_1= 0.4, 0.5 and X3 for ρ_1= 0.5 it is less by 0.002, 0.003 and 0.001, respectively. The estimated sizes of the LM1 test for the 100 per cent of the cases in table 6.6 and 6.7 differ significantly than the five per cent nominal size for the sample sizes considered in this study.

---

[8] Their subroutine deals with autocorrelated errors, whilst here it is concerned withequicorrelated disturbances for the 3SLR model.

The sizes of the LM1 test have the tendency to increase with an increase in the sample size, whereas the sizes of the LM2 test have a tendency to decrease. Generally, the sizes remain less than the five percent nominal level. The lowest sizes of the LM1 test for $X1$, $n=24$ and $n=48$ are 0.013 and 0.019 whereas for $X2$, $n=24$ and $n=64$ they are 0.010 and 0.017, respectively. While the lowest estimated sizes for $X3$, $n=24$ and $n=72$ are 0.018 and 0.023, the highest estimated sizes of the LM1 test for $X1$, $n=24$ and $n=48$ are 0.018 and 0.023, for $X2$, $n=24$ and $n=64$ are 0.013 and 0.019, whilst for $X3$, $n=24$ and $n=72$ they are 0.023 and 0.028, respectively. Thus the sizes of the LM1 test increases as the sample sizes increases, whilst the sizes of the LM2 test behave opposite effect to that of the sizes of the LM1 test.

The rest of this subsection will concentrate on discussing the power behaviour of the two versions of the POI tests namely, $s(0.1, 0.1)$ and $s(0.25, 0.25)$, the LM1 and the LM2 tests. The power calculations of all these tests for the various design matrices, different sample-sizes and different cluster-subcluster-sizes are shown in tables 6.8 to 6.10.

In general, under H_a, the power function of all the tests increases with the increase in the sample size, subcluster size, the number of subclusters in the main cluster, and the values of ρ_1 and ρ_2. The power of the POI tests, for small-sample sizes ($n=24$) are always greater than or equal to the power of the LM1 test. Exceptions occur in table 2.9 for $\rho_1=0.1, 0.2$ and $\rho_2=0.05$ where the power of LM1 test is a bit larger than that of POI tests by 0.002, which is of course not significant. As the sample size increases the power of the LM1 test is becoming closer to the POI tests for the higher values of ρ_1 and ρ_2. While for the extremely lower values of ρ_1, the power of the LM1 test is marginally better than its rival POI test, but this is only just. However, the LM2 test has always the lowest power among all the tests under consideration in this study.

Generally speaking, the power difference between POI tests and the LM1 test is significant for all values of ρ_1, $\rho_2 > 0.2$. The exceptions in table 6.8 are for $\rho_1 = \rho_2 = 0.1$, $\rho_1 = 0.2$, $\rho_2 = 0.1$ and $\rho_1 = 0.0, 0.05, 0.1, 0.2$, with $\rho_2 = 0.2$ and for $\rho_1 = \rho_2 = 0.2$. For the rest of the combination of the values of ρ_1, ρ_2, which are ≤ 0.2, the power of the LM1 and POI tests is approximately similar.

Finally, we turn to the comparison of powers of the two versions of the POI tests, namely the $s(0.1, 0.1)$ and $s(0.25, 0.25)$ tests of H_0 against H_a. All calculated powers of these tests for $X1$, $X2$ and $X3$ design matrices are prescribed in the first two rows of the tables 6.8 to 6.10 of appendix 6B. These tables reveal that for small-samples (i.e., $n=24$) in tables 6.8 to 6.10 the maximum power difference between these tests ranges from 0.001 to 0.035. For such a case it seems that the $s(0.25, 0.25)$ test has the best overall power properties. For moderate sample sizes, the powers of the $s(0.1, 0.1)$ and $s(0.25, 0.25)$ tests are approximately similar, with the maximum power difference between both tests is 0.003. Therefore, this section concludes that for moderate sample sizes the POI test is approximately UMPI along $\rho_1 = \rho_2$, as its power curve is approximately unchanged for the different values of the parameters under the alternative parameter space.

6.8. THE MSLR MODEL AND THE TESTS

The Model

This section extends the 3SLR model considered in an earlier section to a situation where observations are obtained from a multi-stage cluster design. In this case, the ith clusters (say, states) is divided into subclusters (districts), sub-subclusters (suburbs), sub-sub-subclusters (major street blocks) and so on. Thus, it obtains a total of n observations which are sampled from m first-stage states, with $m(i)$ districts from the ith state, with $m(i,j)$ suburbs from the jth district of the ith state, with $m(i,j,k)$ major street blocks from the kth suburbs of the jth district of the ith state and so on, with $m(i,j,k,l,...,q,r)$ observations from the qth household of the lth street block, of the kth suburb of the jth district of the ith state, such that the sample size is

$$n = \sum_{i=1}^{m} \sum_{j=1}^{m(i)} \sum_{k=1}^{m(i,j)} \cdots \sum_{r=1}^{m(i,j,...,p')} m(i,j,k,...,p',q).$$

Thus, the multistage linear regression model at the rth level can be expressed as

$$y_{ijk...p'qr} = \sum_{\ell=1}^{p} \beta_\ell x_{(ijk...p'qr)\ell} + u_{ijk\cdots p'qr} \qquad (6.31)$$

for observations $r = 1,2,...,m(i,j,k,...,q)$ from the qth household, households $q=1,2,...,m(i,j,...,p'),...,$ from the kth suburb, suburbs $k = 1,2,...,m(i,j)$, from the

jth district, districts $j =1,2,...,m(i)$, from the ith state for states $i =1,...,m$ with $y_{ijk...r}$ as the dependent variable and p independent variables $x_{(ijk...p'q)\ell}$ ℓ $=1,...,p$, one of which may be a constant. The error term $u_{ijk...r}$ can be written as

$$u_{ijk...qr} = v_i + v_{ij} + v_{ijk} + ... + v_{ijk...qr},$$

where v_i is the state (cluster) effects, v_{ij} is the jth district effect in the ith state, v_{ijk} is the kth suburb effect in the jth district of the ith state and so on. The $v_{ijk...qr}$ is the remaining random effect, with these r components of $u_{ijk...qr}$ assumed to be mutually independent and normally distributed with

$$E(v_i) = E(v_{ij}) = ... = E(v_{ijk....qr}) = 0 \tag{6.32}$$

and

$$\mathrm{var}(v_i) = \sigma_1^2, \mathrm{var}(v_{ij}) = \sigma_2^2,..., \mathrm{var}(v_{ijk...qr}) = \sigma_r^2,$$

so that $E(u_{ijk....qr}) = 0$ and

$$
E\left(\begin{smallmatrix} u_{ijk...qr} \\ u_{i'j'k'..q'r'} \end{smallmatrix}\right) = \begin{cases}
0 & \text{for } i \neq i' \text{ any } j, j', k, k', ..., q, q', r, r' \\
\sigma_1^2 & \text{for } i = i' \; j \neq j', \text{ any } k, k', \ell, \ell', ...q, q'r, r' \\
\sigma_1^2 + \sigma_2^2 & \text{for } i = i', j = j', k \neq k', \text{any } \ell, \ell...q, q', r, r' \\
\sigma_1^2 + \sigma_2^2 + \sigma_3^2 & \text{for } i = i', j = j', k = k', \ell = \ell, ...\text{any } q, q', r, r' \\
\quad \vdots & \\
\sigma_1^2 + \sigma_2^2 + \sigma_3^2 + ... + \sigma_{r-1}^2 & \text{for } (i, j, k, .., q) = (i', j', k', .., q'), r \neq r' \\
\sigma_1^2 + \sigma_2^2 + ... + \sigma_{r-1}^2 + \sigma_r^2 & \text{for } (i, j, k, ..., q) = (i', j', .., q', r')
\end{cases}
\tag{6.33}
$$

for $r,r'=1,2,...,m(i,j,k,...,q)$

$q,q'=1,2,...,m(i,j,k,...,p')$

$k,k'=1,2,...,m(i,j)$

$j,j'=1,2,...,m(i)$

$i,i'=1,2,....,m.$

If $\sigma_1^2 + \sigma_2^2 + ... + \sigma_\tau^2 = \sigma^2$, then the coefficient of intra-cluster or state equicorrelation is defined as

$$\rho_1 = \sigma_1^2 / \sigma^2.$$

The equicorrelation coefficient of intra-district or subcluster and the intra-suburb are

$$\rho_2 = \frac{\sigma_1^2 + \sigma_2^2}{\sigma^2} \text{ and } \rho_3 = \frac{\sigma_1^2 + \sigma_2^2 + \sigma_3^2}{\sigma^2},$$

respectively, and so on.

Generally, the jth intra-subdivision equicorrelation is defined as

$$\rho_j' = \sum_{i=1}^{j'} \sigma_i^2 / \sigma^2, \text{ for } j' = 1,...,(\tau-1). \tag{6.34}$$

The regression model (6.31) under (6.32), (6.33) and (6.34) can be written in matrix form as

$$y = X\beta + u \tag{6.35}$$

in which y and u are $n \times 1$, β is $p \times 1$ and

$$u \sim N\left(0, \sigma^2 \Omega(\rho_1, \rho_2, ..., \rho_{\tau-1})\right),$$

where

$$\Omega(\rho_1, \rho_2, ..., \rho_{\tau-1}) = \bigoplus_{i=1}^{m} \Omega_i(\rho_1, \rho_2, ..., \rho_{\tau-1}), \tag{6.36}$$

cluster diagonal with submatrices Ω_i. For simplicity of this study, only the case of balanced data will be considered. If $s_1 = m(i)$, $s_2 = m(i,j)$, $s_3 = m(i,j,k)$,..., $s_{p'} = m(i,j,k,...,p')$ $s_q = m(i,j,k,...,1)$. Then (6.36) can be simplified as

$$\Omega(\rho_1,...,\rho_{\tau-1}) = \left(1 - \sum_{i=1}^{\tau-1} \rho_i\right)I_n + \sum_{i=1}^{\tau-1} \rho_i D_i$$

(6.37)

where

$$D_1 = I_m \otimes E_{\left(s_1 s_2 \cdots s_q\right)}$$

$$D_2 = I_{ms_1} \otimes E_{\left(s_2 s_3 \cdots s_q\right)}$$

$$D_3 = I_{ms_1 s_2} \otimes E_{\left(s_3 s_4 \cdots s_q\right)}$$

$$\vdots \qquad \vdots$$

$$D_q = I_{\left(ms_1 s_2 \cdots s_q\right)} \otimes E_{sq}.$$

(6.38)

The Testing Problem and the Tests

This subsection will test the simple null hypothesis

$$H_0 : \rho_i = 0, \text{ against}$$

(6.39)

$$H_a : \rho_i > 0, \text{ for } i = 1,...,\tau - 1, \left(\text{excluding } H_0\right).$$

(6.40)

Note testing the general form of (6.9) against (6.10) is invariant under the general group of transformations of the form (6.11), with maximal invariant new multi-dimensional vector ϑ, given by (6.12). The probability density function of ϑ in this case, under (6.36) and (6.37) can be written in the form of (6.13) as

$$f(\vartheta,\rho_1,...,\rho_{\tau-1}) = 1/2\Gamma\left(\frac{g}{2}\right)\pi^{-g/2}\left|P\Omega(\rho_1,...,\rho_{\tau-1})P'\right|^{-1/2}$$

$$\left(\vartheta'(P\Omega(\rho_1,...,\rho_{\tau-1})P')\vartheta\right)^{-g/2} d\vartheta, \qquad (6.41)$$

where $d\vartheta$ denotes the uniform measure on the unit sphere, and $g=n-p$, $M_x=I_n-X(X'X)^{-1}X'$ and P is a $g\times n$ matrix such that $PP'=I_g$ and $P'P=M_x$. Note that $\rho_1,\rho_2,...,\rho_{\tau}-1$ are the unknown parameters in (6.41).

Point Optimal Invariant Test

This subsection chooses $(\rho_1,\rho_2,...,\rho(\tau-1))'=(\rho_{11}...,\rho_{21},\rho(\tau-1)),1)'$, a point in the alternative parameter space. Based on the maximal invariant (6.12), with the density (6.41), one can construct the POI test for testing (6.39) against (6.40), that is to reject H_0 for small values of

$$\bar{s} = \frac{\hat{u}'\Delta\hat{u}}{\hat{u}'\hat{u}} \qquad (6.42)$$

where

$$\Delta = \Omega_1^{-1} - \Omega_1^{-1}X\left(X\Omega_1^{-1}X\right)^{-1}X\Omega_1^{-1}$$

$$\Omega_1 = \Omega\left(\rho_{11},\rho_{21},...,\rho_{(\tau-1)1}\right)$$

and $\hat{u}$ is the OLS residual vector, $\hat{u} = M_x y$. Thus a test based on $\bar{s}$ gives the maximum attainable power at $\left(\rho_{11},\rho_{21},...,\rho_{(\tau-1)1}\right)$ in the class of invariant tests. Furthermore, like s in (6.17), $\bar{s}$ in (6.42) is also a ratio of quadratic forms in multi-normal variates and therefore its critical values and power function can be solved by using the standard numerical algorithm, which is mentioned in earlier chapters of this book.

Locally Most Mean Powerful Invariant Test

Based on the maximal invariant vector of the form (6.12), when more than one parameter is being tested, one can apply the usual theorem to develop a LMMPI test for testing (6.39) against (6.40). This is to reject H_0, i.e. (6.39), for small values of

$$\bar{\bar{s}} = \frac{\hat{u}'A\hat{u}}{\hat{u}'\hat{u}},$$

where

$$A = (\tau - 1)I_n - \sum_{i=1}^{\tau-1} D_i,$$

which is a generalized form of A, in (6.15). Here D_i for $i = 1,2,...,(\tau\text{-}1)$, is already been defined by (6.38). If one wishes, one can also write (6.43) in a simplified form like that of (6.16), with the critical values and the power of this test also calculated using the standard numerical techniques mentioned earlier in the book.

6.8. CONCLUDING REMARKS

In this chapter the problem of testing for equicorrelation in linear regression errors due to multi-cluster effects was developed. The 3SLR model was considered, a POI and LMMPI test for testing cluster/subcluster effects for this model were constructed. The power study reported in the section entitled 'Power comparison of the POI and the LMMPI tests' shows that the LMMPI test has good power for the data sets used, and therefore provides support for its usefulness in developing these tests for testing multi-cluster effect discussed in the previous section. The second issue considered in this chapter was the testing for subcluster effects in the presence of main cluster effects. A POI test and the asymptotic LM tests for this complicated problem was developed, with the exact sizes and powers of the two versions of the POI test computed using the method illustrated in the section entitled 'Testing $\rho 2$ in the presence of $\rho 1$', whilst those of the asymptotic tests, namely LM1 and LM2 were estimated using the Monte Carlo simulation method.

The results obtained from detailed calculations reported in appendix 6B revealed that in most of the cases, estimated sizes of the LM1 test are less than or equal to the nominal size of 0.05. The sizes of the LM1 test have the tendency to increase with sample size, and it was found that the POI test is marginally better than the LM1 test for small and moderate sample sizes. It was also found that the POI test is approximately UMPI, and that based on these encouraging power results the POI and the LMMPI tests for testing multi-cluster effects while dealing with the multi-stage linear regression model, have been generalised. Furthermore, if one wishes, one can extend the testing problem of the section entitled 'Testing $\rho 2$ in the presence of $\rho 1$' to a more general form of testing lower level cluster effects in the presence of higher level cluster effects.

APPENDIX 6A: THE DATA USED IN THIS CHAPTER

The symbols used here are described on page 137 of chapter 6, where $i=1,2,...,m$, refers to the number of clusters, $j=1,2,...,N_i=s$, refer to the number of subclusters in the ith cluster, and $k=1,2,...,T$, refers to the number of observations in the jth subcluster of the ith cluster.

The data selected in table 6.1 is based on a multistage cluster sampling design for the year 1986-87. The primary sampling units or first-stage are the two main administrative regions of Bangladesh (Khulna and Rajshahi divisions), from which four and three districts (subclusters) are randomly selected. From each district a number of farms, proportional to the cultivated area of the district are selected. There are 600 farms selected from Khulna and Rajshahi divisions, with the district wise farm distribution of each division is given in table 6.2.

Table 6.1. X1: Bangladesh agricultural data for n=96, m=2, s=3, T=16

i	j	k	x_{ijk2}	x_{ijk3}	i	j	k	x_{ijk2}	x_{ijk3}
1	1	1	2.0	13.0	1	2	9	3.8	14.7
1	1	2	5.0	14.0	1	2	10	4.9	14.6
1	1	3	1.9	13.0	1	2	11	7.0	13.0
1	1	4	2.3	12.8	1	2	12	8.5	12.1
1	1	5	7.0	13.5	1	2	13	6.6	13.8
1	1	6	6.5	14.0	1	2	14	8.0	12.6
1	1	7	4.0	14.5	1	2	15	11.2	12.0
1	1	8	3.8	14.6	1	2	16	9.3	11.9
1	1	9	4.5	14.4					
1	1	10	5.0	14.8	1	3	1	2.0	13.2
1	1	11	9.0	12.2	1	3	2	5.0	14.5
1	1	12	11.0	12.0	1	3	3	3.8	14.0
1	1	13	15.0	11.0	1	3	4	4.0	14.3
1	1	14	6.0	14.0	1	3	5	5.5	14.2
1	1	15	9.0	12.0	1	3	6	4.7	14.6
1	1	16	12.0	11.5	1	3	7	6.8	13.8
					1	3	8	2.5	13.6
1	2	1	1.9	13.0	1	3	9	3.9	14.0
1	2	2	5.0	14.2	1	3	10	5.0	14.6
1	2	3	3.7	14.6	1	3	11	6.9	13.8
1	2	4	4.0	14.8	1	3	12	8.4	12.9
1	2	5	21.6	10.3	1	3	13	6.6	13.9
1	2	6	4.6	14.3	1	3	14	8.0	13.0
1	2	7	7.8	12.7	1	3	15	10.5	12.6
1	2	8	2.5	13.9	1	3	16	9.4	12.8
2	1	1	4.5	14.8	2	2	9	2.8	13.7
2	1	2	3.9	14.4	2	2	10	4.0	14.2
2	1	3	5.0	15.0	2	2	11	6.5	14.5
2	1	4	6.0	15.1	2	2	12	8.0	14.1
2	1	5	3.7	14.4	2	2	13	7.0	14.0
2	1	6	4.0	14.5	2	2	14	9.2	13.8
2	1	7	8.0	14.7	2	2	15	5.5	14.4
2	1	8	9.5	14.3	2	2	16	10.0	13.9
2	1	9	11.0	14.2					
2	1	10	10.0	14.0	2	3	1	1.9	13.9
2	1	11	6.5	15.0	2	3	2	3.0	14.2
2	1	12	2.9	14.2	2	3	3	2.5	14.0
2	1	13	7.7	14.8	2	3	4	4.0	14.4
2	1	14	12.0	13.9	2	3	5	5.2	15.0
2	1	15	8.5	14.7	2	3	6	7.0	14.8
2	1	16	1.9	14.0	2	3	7	4.6	14.9
					2	3	8	6.0	15.2
2	2	1	2.5	13.7	2	3	9	5.5	15.2
2	2	2	4.0	14.1	2	3	10	7.0	15.0
2	2	3	1.7	13.4	2	3	11	2.9	14.2
2	2	4	3.7	14.0	2	3	12	5.0	15.0
2	2	5	5.0	14.5	2	3	13	8.0	14.5
2	2	6	4.3	14.2	2	3	14	6.4	15.1
2	2	7	1.9	13.5	2	3	15	4.8	14.9
2	2	8	5.3	14.3	2	3	16	11.0	14.2

Table 6.2. The sampling design and the district-wise farm distribution

Division/District	Number of farms
Khulna	
Jessore	78
Khulna	92
Kushtia	79
Satkhira	80
Subtotal	329
Rajshahi	
Natore	87
Nawabgunj	85
Rajshahi	99
Subtotal	271
Total	600

However, in this book, for computational ease and brevity, the exact multistage cluster sampling procedure has been deviated. Here a subsample of 96 farms from seven selected districts of both divisions has been taken, with the measurements of the variables used in this data are as follows:

a) x_{ijk2}: Biological-Chemical (BC) input:

This concludes both high yielding variety (HYV) and local variety of seeds, the chemical fertilizers, pesticides and insecticides. BC input per acre is measured in terms of money, i.e. in thousands of Taka.

b) x_{ijk3}: Human labour:

The data on human labour is given in adult man-days, which includes family as well as hired labour. In order to compute the wage-bill, family labour is assigned imputed value which is equal to the average of wages of casual hired labour and permanently hired labour.

Table 6.3. X2: Artificially generated data for $n=96$, $m=2$, $s=2$, $T=24$

I	j	k	x_{ijk2}	x_{ijk3}	x_{ijk4}
1	1	1	102.44028	14.62891	14.32892
1	1	2	17.69850	38.22080	4.26574
1	1	3	45.39083	12.41084	10.72648
1	1	4	22.10381	15.73724	7.26575
1	1	5	86.54461	11.51338	18.44165
1	1	6	83.43267	15.56228	6.41979
1	1	7	6.06479	14.29658	8.27100
1	1	8	30.97777	8.77406	9.70438
1	1	9	29.61556	17.14609	12.08422
1	1	10	20.02457	18.50295	4.89055
1	1	11	19.88290	20.42256	5.26336
1	1	12	7.97451	4.22572	14.84995
1	1	13	193.92471	66.66033	11.02774
1	1	14	70.74133	44.00222	14.97263
1	1	15	24.13951	16.04665	4.72805
1	1	16	21.09276	10.30161	1.68796
1	1	17	23.76728	13.89932	5.78294
1	1	18	2.47052	27.68553	1.99841
1	1	19	20.71276	56.91961	8.11510
1	1	20	2.79513	22.05987	1.57264
1	1	21	102.08237	199.58296	0.47101
1	1	22	6.71692	22.43248	12.04968
1	1	23	71.28814	59.62470	19.62987
1	1	24	36.30734	40.63420	15.53035
1	2	1	2.43445	38.22654	5.97177
1	2	2	15.77600	4.90910	4.52484
1	2	3	19.44831	36.81002	5.93403
1	2	4	4.36060	22.74602	17.63734
1	2	5	65.02488	14.40334	13.28677
1	2	6	42.18986	27.28044	3.74893
1	2	7	16.43922	19.17658	14.65504
1	2	8	6.06940	34.57428	9.15907
1	2	9	49.92091	23.44600	8.05458
1	2	10	3.18748	8.62565	1.81427
1	2	11	3.55431	30.22713	10.14179
1	2	12	10.69643	37.75961	3.13892
1	2	13	2.69417	6.92328	2.15610
1	2	14	13.33684	38.13948	19.83900
1	2	15	19.46718	82.74967	19.90378
1	2	16	6.42926	16.27630	14.12346
1	2	17	56.03015	22.88885	13.34708
1	2	18	8.53581	22.77424	9.83104
1	2	19	17.06023	17.07747	0.17146
1	2	20	22.57290	29.96200	2.71205
1	2	21	0.66291	30.19209	18.85556
1	2	22	40.69153	159.69408	14.34513
1	2	23	29.88689	3.33813	4.01511
1	2	24	79.88276	7.95492	19.61812
2	1	1	32.10366	25.13421	3.88157
2	1	2	65.93636	74.07070	16.12026

I	j	k	x_{ijk2}	x_{ijk3}	x_{ijk4}
2	1	3	20.40541	19.44403	10.45981
2	1	4	4.70022	18.24024	8.92963
2	1	5	50.39289	21.80480	0.93681
2	1	6	16.15953	3.48069	4.88509
2	1	7	23.26128	5.70259	8.03407
2	1	8	30.43460	15.06313	5.25635
2	1	9	49.90344	2.17226	10.96155
2	1	10	9.07107	24.08743	3.51882
2	1	11	4.87983	12.93944	1.18945
2	1	12	51.13453	10.00375	14.45123
2	1	13	72.33007	37.02710	12.15640
2	1	14	12.46095	28.49646	10.75508
2	1	15	21.89569	33.75606	3.02889
2	1	16	10.86493	12.99299	2.59360
2	1	17	11.08796	26.97580	17.46648
2	1	18	13.62096	55.59037	12.61626
2	1	19	7.54104	15.10099	12.78103
2	1	20	55.03338	6.18224	12.92152
2	1	21	30.15136	65.37304	16.65537
2	1	22	11.66532	41.10583	9.76576
2	1	23	8.85764	30.37288	11.45818
2	1	24	15.40990	23.15244	5.40089
2	2	1	9.97408	6.71242	13.90643
2	2	2	21.89000	7.84958	2.96579
2	2	3	54.53267	3.00991	4.13541
2	2	4	34.41491	4.58495	8.61873
2	2	5	32.83451	10.64361	6.88896
2	2	6	53.37914	7.32608	13.82809
2	2	7	30.39810	38.58833	12.28813
2	2	8	7.13028	11.13474	8.68824
2	2	9	55.80090	19.55713	8.22674
2	2	10	74.26427	20.56931	13.04514
2	2	11	62.25320	4.83799	14.62588
2	2	12	10.83466	19.09755	14.85305
2	2	13	16.75522	49.45385	5.30633
2	2	14	24.59812	89.69814	3.28932
2	2	15	6.45077	71.45801	9.85965
2	2	16	25.48525	25.43230	16.47815
2	2	17	2.57735	27.98616	9.36280
2	2	18	7.29960	6.18768	18.96755
2	2	19	36.23263	26.15473	9.84032
2	2	20	9.96431	95.91381	1.11511
2	2	21	17.41201	2.03546	19.78022
2	2	22	31.11064	15.47893	0.01891
2	2	23	329.25746	240.14670	6.50564
2	2	24	6.74175	39.45260	18.32898

M. Ishaq Bhatti

APPENDIX 6B. TABULATION OF RESULTS OF THE COMPARATIVE STUDY

Table 6.4. Powers of the LMMPI test for testing
$$H_0: \rho_1 = \rho_2 = 0, \text{ against } H_a: \rho_1 > 0, \rho_2 > 0$$
(Figures in brackets are point optimal powers)

Case X1a, NixT = 2x12

0.4	0.736	0.775	0.810	0.868	0.954
0.2	0.446	0.514	0.572	0.668	0.813
	(0.498)			(0.695)	
0.1	0.240	0.323	0.395	0.511	0.684
0.05	0.135	0.220	0.297	0.422	0.604
0.0	0.050	0.128	0.206	0.337	0.525
				(0.361)	
$\rho2/\rho1$	0.0	0.05	0.1	0.2	0.4

Case X1b, NixT = 3x8

0.4	0.739	0.786	0.826	0.891	0.974
0.2	0.396	0.473	0.539	0.648	0.815
	(0.492)			(0.700)	
0.1	0.197	0.279	0.353	0.475	0.660
0.05	0.112	0.191	0.265	0.390	0.577
0.0	0.050	0.119	0.190	0.316	0.502
				(0.341)	
$\rho2/\rho1$	0.0	0.05	0.1	0.2	0.4

Case X1c, NixT = 2x24

0.4	0.892	0.912	0.928	0.953	0.985
0.2	0.691	0.749	0.791	0.851	0.926
	(0.738)			(0.873)	
0.1	0.444	0.556	0.630	0.729	0.847
0.05	0.250	0.406	0.506	0.632	0.775
0.0	0.050	0.235	0.364	0.519	0.683
				(0.535)	
$\rho2/\rho1$	0.0	0.05	0.1	0.2	0.4

Case X1d, NixT = 3x16

0.4	0.926	0.943	0.957	0.976	0.996
0.2	0.691	0.754	0.800	0.865	0.945
	(0.758)			(0.894)	
0.1	0.402	0.513	0.593	0.703	0.841
0.05	0.209	0.345	0.444	0.579	0.742

| 0.0 | 0.050 | 0.186 | 0.297 | 0.450
(0.478) | 0.628 |
| $\rho 2/\rho 1$ | 0.0 | 0.05 | 0.1 | 0.2 | 0.4 |

Case X2a, NixT = 2x12

0.4	0.707	0.753	0.791	0.854	0.947
0.2	0.425	0.508	0.573	0.672	0.811
	(0.526)			(0.735)	
0.1	0.235	0.347	0.432	0.554	0.713
0.05	0.136	0.262	0.358	0.494	0.662
0.0	0.050	0.178	0.286	0.436	0.616
				(0.443)	
$\rho 2/\rho 1$	0.0	0.05	0.1	0.2	0.4

Case X2b, NixT = 2x12

0.4	0.646	0.712	0.764	0.847	0.958
0.2	0.332	0.439	0.518	0.635	0.794
	(0.511)			(0.747)	
0.1	0.176	0.302	0.397	0.529	0.696
0.05	0.107	0.239	0.340	0.481	0.654
0.0	0.050	0.179	0.286	0.437	0.616
				(0.443)	
$\rho 2/\rho 1$	0.0	0.05	0.1	0.2	0.4

Table 6.5.Calculated values of ρ_{10}^{*} and c_{α}^{*} for the POI-tests s(0.1,0.1), s(0.25,0.25) and c_{α} for the LM1 and LM2 tests at the five percent significance level for selected value of n, m, s and T

Design Matrix	s(0.1, 0.1) test		s(o.25, 0.25) test		LM1 test	LM2 test	n	M	s	T
	ρ_{10}^{*}	c_{α}^{*}	ρ_{10}^{*}	c_{α}^{*}	c_{α}	c_{α}				
X1(nx3)	0.1279	0.9816	0.3559	1.0786	1.4947	3.5528	48	2	3	8
X1(nx3)	0.1252	0.9591	0.3298	0.9924	1.2966	3.9919	24	2	3	4
X2(nx3)	0.1222	0.9964	0.3441	1.1103	1.2743	3.7645	64	2	4	8
X2(nx3)	0.1208	0.9691	0.3232	1.0112	1.0655	3.8244	24	2	3	4
X3(nx3)	0.1313	0.9987	0.3693	1.1139	1.5421	3.6574	72	2	3	12
X3(nx3)	0.1192	0.9579	0.3175	0.9950	1.4068	4.0323	24	2	3	4

Table 6.6. Estimated sizes of LM1 and LM2 tests for testing ρ_2 in the presence of ρ_1 using asymptotic critical values at 5% nominal level, for $n=24$ and $p=3$

Data Matrices	ρ_1	Test Statistics	
		LM1*	LM2
X1	0.0	.015	.059
	0.05	.018	.055
	0.1	.016	.050
	0.2	.016	.047
	0.3	.016	.038*
	0.4	.014	.031*
	0.5	.013	.026*
X2	0.0	.013	.048
	0.05	.012	.043
	0.1	.013	.040*
	0.2	.012	.032*
	0.3	.012	.030*
	0.4	.012	.028*
	0.5	.010	.024*
X3	0.0	.021	.059
	0.05	.023	.050
	0.1	.023	.045
	0.2	.022	.038*
	0.3	.019	.033*
	0.4	.019	.032*
	0.5	.018	.031*

* Indicates the sizes are significantly different than the nominal size of five percent.

Table 6.7. Estimated sizes of LM1 and LM2 tests for testing ρ_2 in the presence of ρ_1 using asymptotic critical values at 5% nominal level, for $p=3$

Data Matrices	Sample sizes	ρ_1	Test Statistics	
			LM1*	LM2*
$X1$	48	0.0	.019	.033
$(n\times3)$		0.05	.020	.029
		0.1	.021	.031
		0.2	.021	.029
		0.3	.022	.028
		0.4	.023	.028
		0.5	.023	.027
$X2$	64	0.0	.019	.047**
$(n\times3)$		0.05	.018	.038
		0.1	.018	.032
		0.2	.016	.028
		0.3	.016	.025
		0.4	.016	.023
		0.5	.017	.022
$X3$	72	0.0	.025	.042**
$(n\times3)$		0.05	.026	.034
		0.1	.025	.030
		0.2	.028	.033
		0.3	.025	.030
		0.4	.025	.028
		0.5	.023	.024

** Indicates exceptional cases, when the size of the LM2 test differs insignificantly from the nominal size of five percent.

Table 6.8. Calculated values of the power function for $X1$ with $n=24$, $m=2$, $s=3$, $T=4$ and $p=3$, using five percent critical values for testing H_0: $\rho_2 = 0$, $\rho_1 > 0$, against H_a: $\rho_1 > 0$, $\rho_2 > 0$

Tests $\rho_2 \backslash \rho_1 =$		0.0	0.05	0.1	0.2	0.3	0.4	0.5
s(0.1,0.1)	0.0	.050	.050	.050	.050	.050	.050	.050
s(0.25,0.25)		.050	.050	.050	.050	.050	.050	.050
LM1		.043	.048	.050	.049	.047	.043	.041*
LM2		.050	.045	.041	.038	.033	.026	.024
s(0.1,0.1)	0.05	.081	.082	.084	.089	.095	.103	.115
s(0.25,0.25)		.081	.082	.084	.089	.095	.104	.117
LM1		.074	.081	.080	.080	.082*	.086*	.095*
LM2		.049	.050	.047	.043	.044	.043	.038
s(0.1,0.1)	0.1	.119	.123	.128	.140	.155	.177	.208
s(0.25,0.25)		.118	.122	.127	.140	.156	.179	.213
LM1		.109	.112	.112*	.127*	.142*	.160*	.180*
LM2		.056	.056	.056	.056	.061	.067	.078
s(0.1,0.1)	0.2	.217	.228	.241	.271	.311	.365	.442
s(0.25,0.25)		.216	.227	.240	.272	.315	.374	.459
LM1		.198*	.201*	.217*	.247*	.277*	.311*	.370*
LM2		.100	.104	.110	.129	.147	.171	.217
s(0.1,0.1)	0.3	.335	.354	.375	.426	.490	.574	.683
s(0.25,0.25)		.335	.354	.377	.431	.501	.593	.714
LM1		.308*	.318*	.337*	.384*	.440*	.510*	.593*
LM2		.182	.190	.204	.241	.277	.325	.420
s(0.1,0.1)	0.4	.464	.490	.519	.586	.668	.766	.876
s(0.25,0.25)		.465	.493	.524	.596	.685	.792	.911
LM1		.434*	.458*	.484*	.537*	.607*	.684*	.788*
LM2		.270	.292	.315	.369	.432	.529	.646
s(0.1,0.1)	0.5	.594	.625	.659	.735	.821	.910	.976
s(0.25,0.25)		.598	.631	.668	.750	.843	.937	.998
LM1		.565*	.587*	.619*	.680*	.751*	.843*	.932*
LM2		.391	.416	.448	.523	.615	.722	.853

*Indicates the difference in power between POI and LM1 test is significant, whereas the values without * shows the difference in power between POI and LM1 test is not significant.

7. TESTS FOR THE SNSG MODEL

7.1. INTRODUCTION[1]

It is common in business surveys, opinion polls and in other sorts of routine socio-economic based questionnaires that businessmen and sociologists always deal with quite a Large Number of Small Groups or Clusters (LNSG) of observations. In these surveys the observations for each group are obtained under similar conditions and are therefore correlated. For example, for routine household surveys, duplicate and/or triplicate samples from a locality might be repeatedly taken, or in hospitals during a large study blood pressure measurements might be taken several times from the same patient. Under such situations there may exist of course an autocorrelation within each group.

Cox and Solomon (1988) discussed the problem of testing for such an autocorrelation coefficient in large numbers of small samples. There study was particularly concerned with group sample sizes of three and illustrated a suggested procedure by applying it to triple observations of pulse rates for 200 people. This chapter re-examines Cox and Solomon's problem of testing for a non-zero autocorrelation coefficient, ρ, when there is a number of very small samples LNSG.

The aim of this chapter is to see if one can succeed in constructing some optimal tests (which have been discussed in previous chapters) for cases where there are m independent samples, each of size k. This chapter applies previously developed tests like LBI, POI and LMMPI tests for testing ρ in the

---

[1] Part of the findings reported in this chapter appeared in Bhatti (1992).

LSNG-model. The plan of this chapter is as follows. In the following section, a general model for m independent samples of size k is given, with LBI, POI and LMMPI tests derived. The final section of this chapter will contain some concluding remarks.

7.2. THE LNSG MODEL AND THE TESTS

The LNSG model

Suppose there are m independent samples of k observations, which are denoted by

$$y_{it}, t = 1,...,k,; i = 1,...,m;$$

in which each sample has unknown mean

$$E(y_{it}) = \mu_i, i = 1,...,m.$$

Let $u_{it} = y_{it} - \mu_i$ denote deviations about these means. It is assumed that u_{it} follows the stationary AR(1) process

$$u_{it} = \rho u_{it-1} + \varepsilon_{it}$$

where

$$\varepsilon_{it} \sim IN(0,\sigma^2),$$

so that

$$u(i) = (u_{i1},...,u_{ik})' \sim N(0,\sigma^2 \Sigma(\rho))$$

where

$\Sigma(\rho)$ is the $k \times k$ matrix

$$\Sigma(\rho) = \frac{1}{1-\rho^2} \begin{bmatrix} 1 & \rho & \rho^2 & \cdots & \rho^{k-1} \\ \rho & 1 & \rho & & \vdots \\ \rho^2 & \rho & 1 & & \vdots \\ & & & \ddots & \\ \vdots & & & 1 & \rho \\ \rho^{k-1} & \cdots & & \rho & 1 \end{bmatrix}.$$

Thus $y(i) = (y_{i1}, \ldots, y_{ik})' \sim N\left(\mu_i \ell, \sigma^2 \Sigma(\rho)\right)$

where

$$\ell = (1, \ldots, 1)' \text{ or } y(i) = \mu_i \ell + u(i), i = 1, \ldots, m, \qquad (7.1)$$

in which

$$u(i) \sim N\left(0, \sigma^2 \Sigma(\rho)\right).$$

If $n = mk$ is the total number of observations, and if we stack the $y(i)$ and $u(i)$ to form the $n \times 1$ vectors

$$y = (y(1)', \ldots, y(m)')' \text{ and } u = (u(1)', \ldots, u(m)')',$$

respectively, then

$$y = X\mu + u, \qquad (7.2)$$

where X is the $n \times m$ matrix of the form, $X = I_m \otimes \ell$, $\otimes$, the Kroneker product and $\mu = (\mu_1, \ldots, \mu_m)'$. Therefore

$$u \sim N\left(0, \sigma^2 \Delta(\rho)\right) \qquad (7.3)$$

in which $\Delta(\rho) = I_m \otimes \Sigma(\rho)$ is an $n \times n$ matrix. In what follows, model (7.1) and its matrix version (7.2) can be extended to include regressors if required without loss of generality.

Optimal Testing for LNSG

The main problem of interest in this section is to test $H_0: \rho = 0$ against $H_a: \rho > 0$ in LNSG model defined in (7.2) under (7.3). Note that this testing problem is invariant under the group of transformations

$$y \rightarrow \gamma_0 y + X\gamma$$

where γ_0 is a positive scalar and γ is a $p \times 1$ vector. Again the three similar optimal tests are constructed, namely the LBI test, the POI test and the LMMPI tests, which have been discussed already in detail in previous chapters of this book.

Locally Best Invariant Tests for LNSG

Here the locally best invariant test for the above problem is to reject H_0 for small values of

$$r = \frac{e' A_0 e}{e' e}$$

where

$$A_0 = \frac{-\partial \Delta(\rho)}{\partial \rho} \bigg|_{\rho=0}$$

and e is the vector of OLS residuals from model (7.2), which provides a LBI test against H_a.

Now, one can easily note that $\left. \dfrac{-\partial \Sigma(\rho)}{\partial \rho} \right|_{\rho=0} = A_1$ (say),

where

$$A_1 = \begin{bmatrix} 0 & -1 & 0 \ldots & & 0 \\ -1 & 0 & -1 \ldots & & 0 \\ 0 & -1 & 0 & & \vdots \\ & & & \ddots & \\ \vdots & & & & 0-1 \\ 0 & 0 \ldots & -1 & 0 & 0 \end{bmatrix}, \tag{7.4}$$

so that

$$A_0 = I_m \otimes A_1$$

and

$$r = \left\{ \sum_{i=1}^{m} e(i)' A_1 e(i) \right\} \Big/ e'e \tag{7.5}$$

$$= -2 \sum_{i=1}^{m} \sum_{t=2}^{k} e_{it} e_{it-1} \Big/ \sum_{i=1}^{m} \sum_{t=1}^{k} e_{it}^2 .$$

The form of equation (7.5) is a ratio of quadratic forms in normal variables, with the critical values, c_α, being determined by the usual standard numerical techniques which were used to calculate critical values of the DW statistic (as referred to in earlier chapters). Unfortunately, the distribution of (7.5) is a function of the design matrix X through M_X, where $M_X = I_n - X(X'X)^{-1}X'$. However, bounds for c_α which are independent of X depend on A_0, which can be calculated in an analogous ways to the familiar DW bounds (see King: 1987a, p. 28-29). Methods of approximating the critical values of DW statistics can also be used to approximate c_α (see King: 1987a, p.25-27). If one is interested in testing $H_0 : \rho = 0$, against the two-sided alternative $H_a' : \rho \neq 0$, one can use the generalized Neyman-Pearson lemma (see Ferguson: 1967, p. 238) to obtain a locally best unbiased invariant (LBUI) test, given that the test's size and local unbiased conditions are satisfied.

Point Optimal Invariant Tests for LNSG

This subsection shows how to construct a point optimal invariant test of the null hypothesis $H_0 : \rho = 0$ against an alternative $H_a' : \rho > 0$. To begin, one can consider the simpler problem of testing H_0 against the simple alternative hypothesis $H_1 : \rho = \rho_1 > 0$, in (7.2) and (7.3), where ρ_1 is a known fixed point at which it is required to have optimal power in the alternative parameter space. Using King (1987b, equation 18), the POI test is to reject the null hypothesis for small values of

$$
\begin{aligned}
s(\rho_1) &= \frac{e'Ae}{e'e} \\
&= \frac{u'Au}{u'M_x u}
\end{aligned}
\tag{7.6}
$$

where $e = M_x y = M_x u$,

$$
M_x = I_n - X(X'X)^{-1} X',
$$

$$
A = \Delta^{-1}(\rho_1) - \Delta^{-1}(\rho_1) X \left(X'\Delta^{-1}(\rho_1) X \right)^{-1} X'\Delta^{-1}(\rho_1).
$$

The last equality of (7.6) is obtained by observing that

$$
e'Ae = u'M_x A M_x u = u'Au.
$$

It is important to note that as $s(\rho_1)$ in (7.6), is a ratio of quadratic forms in

normal variables, and hence like for the Durbin-Watson statistic, probabilities
of the form

$$\Pr\left[s\left(\rho_1\right) < c_\alpha\right] = \Pr\left[\frac{u'Au}{u'M_x u} < c_\alpha\right]$$

$$= \Pr\left[u'\left(A - c_\alpha M_x\right)u < 0 \,\middle|\, u \sim N\left(0, \Delta\left(\rho\right)\right)\right] \tag{7.7}$$

$$= \Pr\left[\left(\Delta^{-1/2}\left(\rho_1\right)u\right)'\left(\Delta^{1/2}\left(\rho\right)\right)'\left(\Delta\left(\rho_1\right) - c_\alpha I_n\right)\Delta^{1/2}\left(\rho_1\right)\left(\Delta^{-1/2}\left(\rho_1\right)u\right) < 0\right]$$

can be found by computing

$$\Pr\left[\sum_{i=1}^{n} \lambda_i \xi_i^2 < 0\right], \tag{7.8}$$

where $\lambda_1, \ldots, \lambda_n$ are the eigenvalues o

$$\left(\Delta^{1/2}\left(\rho_1\right)\right)'\left(\Delta\left(\rho_1\right) - c_\alpha M_x\right)\Delta^{1/2}\left(\rho_1\right) \tag{7.9}$$

and $\xi_1^2, \ldots, \xi_n^2$ are independent chi-squared random variables, each with one
degree of freedom.

   To find the exact critical values, (7.8) can be evaluated using standard
GUASS subroutines, as mentioned in earlier chapters. A central question is
however, how should ρ_1 be chosen? Strategies for choosing the point at which
a POI test optimized power is discussed in detail in chapter two, in which one
approach suggested was to chose a ρ_1 value arbitrarily, such as $\rho_1 = 0.3$, -0.5 or
0.75 etc, which was chosen in chapters three and five. Another approach is to
take the limit of $s(\rho_1)$ tests as ρ_1 tends to zero. This approach is called the LBI
test approach, as discussed above, but there seems little point in optimizing
power when it is very low (as the LBI test does) or when it is one or nearly

one. Here Bhatti and King's (1990) approach is highly recommended, in that is optimizing power at a middle power value, say 0.5.

Alternative methods of computing (7.7) without first calculating eigenvalues have been briefly reviewed in chapters five and six. As was noted in chapter five, Palm and Sneek's approach involves using householder transformations to tridiagonalize (7.9), whereas Shively et al's approach suggests the use of modified Kalman filter to calculate (7.7) in $0(n)$ operations. For large sample sizes, this could result in considerable computational savings over the eigenvalue method.

7.3. THE LMMPI TEST FOR THE LNSG MODEL

The error covariance matrix in (7.3) involves only one parameter, ρ, but in practice it may be that the structure of the autocorrelated observations change from sample to sample, or from one group of observations to another. Examples of such groupings would include geographical location, industry, occupation and years of schooling, etc, which might cause the autocorrelation coefficient, ρ, to vary from sample to sample or from group to group. In such cases the covariance matrix $\Delta(\rho)$, in (7.3) can be expressed as

$$\Delta(\bar{\rho}) = \begin{bmatrix} \Sigma(\rho_1) & 0\text{............} & 0 \\ & \ddots & \vdots \\ 0 & \Sigma(\rho_1) & \ddots & \vdots \\ \vdots & \ddots & \ddots & 0 \\ \vdots & & \ddots & \ddots \\ 0\text{............}0 & & \Sigma(\rho_m) \end{bmatrix},$$

$$= \bigoplus_{i=1}^{m} \Sigma(\rho_i), \tag{7.10}$$

which is in a cluster diagonal with submatrices of the form

$$\Sigma(\rho_i) = \left[(1-\rho_i)^2 I_k + \rho_i A_k + \rho_i(1-\rho_i)C_k \right]^{-1},$$

where I_k is the $k \times k$ identity matrix

$$A_k = \begin{bmatrix} 1 & -1 & 0 \dots\dots\dots & & & 0 \\ -1 & 2 & -1 \ddots & & & \vdots \\ 0 & -1 & 2 \ddots \ddots & & & \vdots \\ \vdots & \ddots & \ddots \ddots & & & \vdots \\ \vdots & & \ddots \ddots & -1 & 0 \\ \vdots & & & \ddots & 2 & -1 \\ 0 \dots\dots\dots & & 0 & -1 & & 1 \end{bmatrix}, \quad C_k = \text{diag}(1,0,\dots,1)$$

and

in (7.10), $\rho = (\rho_1, \rho_2, \dots, \rho_m)'$.

The main objective of this subsection is to construct a one-sided test for testing whether these autocorrelation coefficients are equal to zero against the alternative, that they have some positive values, i.e.,

$$H_0 : \rho_1 = \rho_2 = \dots, \rho_m = 0$$

against

$$H_a : \rho_1 \geq 0, \rho_2 \geq 0, \dots, \rho_m \geq 0,$$

with at least one inequality being a strict inequality. This is an m-dimensional hypothesis testing problem, and therefore King and Wu's (1990) theorem can be used in a similar way as in chapter three to construct a LMMPI test for testing H_0 against H_a, which is to reject H_0 for small values of

$$d = \frac{e'Ae}{e'e}, \tag{7.11}$$

where e is the vector of OLS residuals from (7.2) as mentioned earlier and

$$A = \sum_{i=1}^{m} A_i$$

such that from (7.10)

$$A_i = -\left.\frac{\partial \Delta(\bar{\rho})}{\partial \rho_i}\right|_{\rho=0}, i = 1, 2, \ldots, m,$$

(7.11)

$$= \begin{bmatrix}
0 & 0 & \cdots & 0 & \cdots & \cdots & \cdots & \cdots & \cdots & \cdots & \cdots & \cdots & \cdots & \cdots & \cdots & 0 \\
0 & 0 & & \vdots & & & & & & & & & & & & \vdots \\
\vdots & & \ddots & \vdots & & & & & & & & & & & & \vdots \\
0 & \cdots & \cdots & 0 & & & & & & & & & & & & \vdots \\
\vdots & & & & \ddots & & & & & & & & & & & \vdots \\
\vdots & & & & & 0 & -1 & 0 & \cdots & \cdots & 0 & & & & & \vdots \\
\vdots & & & & & -1 & 0 & -1 & & & \vdots & & & & & \vdots \\
\vdots & & & & & & & & \ddots & & \vdots & & & & & \vdots \\
= & \vdots & & & & & 0 & -1 & 0 & & \ddots & \vdots & & & & \vdots \\
\vdots & & & & & \vdots & & & \ddots & & \vdots & & & & & \vdots \\
\vdots & & & & & \vdots & & & & & 0 & -1 & & & & \vdots \\
\vdots & & & & & 0 & \cdots & \cdots & \cdots & -1 & 0 & & & & & \vdots \\
\vdots & & & & & & & & & & & \ddots & & & & \vdots \\
\vdots & & & & & & & & & & & & 0 & 0 & \cdots & 0 \\
\vdots & & & & & & & & & & & & 0 & 0 & & \vdots \\
\vdots & & & & & & & & & & & & \vdots & & \ddots & \vdots \\
0 & \cdots & \cdots & \cdots & \cdots & \cdots & \cdots & \cdots & \cdots & \cdots & \cdots & 0 & \cdots & \cdots & 0
\end{bmatrix}$$

(7.12)

Thus by (7.10) and (7.12), a matrix A can be written as

$$A = I_m \otimes A_1,$$

where A_1 is given by (7.4). It is worth noting that (7.11) is also a LBI test of H_0 in the direction $\rho_1 = \rho_2 = ... = \rho_m > 0$, as it has been noticed in chapters three, four and five.

7.4. CONCLUDING REMARKS

This chapter has developed a precise optimal testing procedure for large numbers of small samples, to detect serial correlation coefficient, ρ, of stationary first order autoregressive process in business and other applied sciences (particularly, statistics, econometrics and biometrics). In the previous section, three optimal tests, i.e. the LBI, POI and LMMPI tests have been derived.

In view of the calculations performed in the previous chapters, especially those in chapters three, four, five and six, with there being a good reason to believe that here also the POI test will perform reasonably well. In fact a number of Monte Carlo studies have already shown that the POI test is typically more powerful than LBI tests (see King: 1985a). From these studies, it can be surmised that POI tests will also be typically more powerful than LMMPI tests. The main interest in this chapter was not so much in the tedious exercise of presenting the proof that POI tests worked yet again, but it was to provide an exercise of building and generalising the Cox and Solomon (1988) model, and then deriving LBI, POI and LMMPI tests for the general model. The next chapter will explore more complex ideas of varying regressors for each cluster.

8. STOCHASTIC MODEL FOR CLUSTER EFFECTS

8.1. INTRODUCTION

In previous chapters we considered 2SLR, 3SLR, LNSG187 and multistage clustered models with fixed regressors that can lead to inappropriate results and misleading conclusions. In practice due to complex structures within and across clusters, variation fix regressor coefficient models are not viable in capturing multi-clustered effects. Economic structures changes or socio-economic and demographic background factors differ from cluster to cluster, with it possible that the response parameters can vary over clusters, groups and/or regions. To illustrate this problem, let us consider few examples. Firstly, equal amounts of labour and capital in a particular production process can yield different levels of output over different clusters of time, in view of technical progress, labour efficiency and managerial ability, which can vary from cluster to cluster. Similarly, identical applications of fertilizer to different clusters of crops or areas of land may yield different levels of output as a result of variations in average temperature, land fertility, rain-fall and agricultural practices. However, Bhatti (2004) pointed out some cautions in comparing estimated results of fixed and random coefficient models.

The main aim of this chapter is to investigate and develop an efficient estimation procedure for the LR model with stochastic and fixed coefficients based on clustered survey data. The focus of our investigation is to estimate cluster effects, ρ, cluster-wise heteroscedasticity and variances of random coefficients, and then summarise optimal tests propositions associated with optimal testing, which has been discussed in previous chapters.

The structure this chapter is as follows. In the subsequent sections, the stochastic coefficient regression model with cluster effects is introduced, and an algorithm for the efficient maximum likelihood estimator (EMLE) is presented. Both random and fix coefficient models are applied using EMLE method in estimating the unknown parameters. In third section, we summarise propositions related to optimal tests, whilst the final section contains some concluding remarks.

8.2. THE MODELS AND THE ESTIMATION PROCEDURES

In chapter five we assumed that n observations are available from a two-stage sample survey data with m clusters. Let m(i) be the number of observations from the ith cluster so that $n = \sum_{i=1}^{m} m(i)$. For ease of notation, here in this chapter we will use a^i to denote ith quantity a associated with the ith cluster while $(a)^i$ will denote quantity a raised to the power i. An exception will be σ^2, which has its usual meaning. Let us express the 2SLR model of (5.1) in the following format[1]

$$y_j^i = \sum_{k=1}^{p} \beta_k x_{jk}^i + u_j^i \qquad (8.1)$$

for observations j=1,2,..., m(i) from clusters i=1,2,...,m, with dependent variables y_j^i and p independent variables, x_{jk}^i, which are assumed non-stochastic and the first of which is a constant. Baltagi (1996) and Bhatti (1999) follow a similar version but with a panel data framework. They both assumed that the regression errors u_j^i are independent between clusters, but equicorrelated within clusters. Hence for (8.1),

---

[1] If we assume $X_{jk}^i = 0$ and $\sigma^2 = 1$, then the model (8.1) becomes a standard symmetric multivariate model (see Bhatti and King (1990) and Bhatti (2000)).

$E(u^i_j) = 0$ for all i and j;

$$E\left(u^i_j u^s_t\right) = \sigma^2 \delta_{is}\left\{\rho + (1-\rho)\delta_{jt}\right\} \qquad (8.2)$$

where ρ is cluster effects, such that $0 \le \rho \le 1, \delta_{ii} = 1$ and $\delta_{ii} = 0, \text{for} i \ne j$. Both Baltagi and Bhatti also assumed that the regression coefficients, β_k; k=1,2,..., p are constant.

The Stochastic Model

In certain applications based on data with a clustering structure, the assumption of response parameter, β_k as fixed may not be appropriate. For example, Hoque (1991) in a similar study with a different dataset used the Langrange multiplier test, (i.e. Rao's (1973) efficient score test, and Wu and Bhatti (1994) PO test in chapter six, for testing the null hypothesis of fixed regression coefficients against an alternative as stochastic. Hoque rejected the null hypothesis in favour of the alternative. Therefore, in this section it may be reasonable to assume that the slope and variances parameters, β_k and σ^2 in (8.1) and (8.2) respectively, change from cluster to cluster (or from subcluster to subcluster). Thus we consider the inter-cluster random coefficients model[2], which regards the coefficients as having constant means and constant variance-covariance's over different clusters, and additionally allows cluster-wise heteroscedasticity in the disturbance terms, u^i_j.

When regression coefficients are varying from one cluster to another, then we can write model (8.1) with inter-cluster random coefficients as,

$$y^i_j = \sum_{k=1}^{p} \beta^i_j x^i_j k + u^i_j, \qquad (8.3)$$

for $j = 1,2,..., m(i)$, cluster size and i=1,2,..., $m = c$, number of clusters. Further, the inter-cluster random coefficients, β^i_j can be decomposed into a

---

[2] Alternatively, known as stationary random coefficient models, e.g. see Hsiao (1986) and Baltagi (1996), which ignore equicorrelation within clusters.

deterministic component (mean) and a random component (deviation from mean). Thus, for the i'th cluster, the regression coefficients can be written as[3]

$$\beta_j^i = \beta_j^i + \varepsilon_j^i; \, k = 2, ..., p, \tag{8.4}$$

where the random component ε_k^i follows the following assumptions,

$$
\left.
\begin{aligned}
&(i). E\left(\varepsilon_k^i\right) = 0, \text{ for all } i \text{ and } k, \\
&(ii). Var\left(\varepsilon_k^i\right) = \sigma_k^2, \, k = 2, ..., p, \\
&(iii). Cov\left(\varepsilon_k^i, \varepsilon_l^j\right) = 0, \text{ for all } i, j, k \text{ and } l \\
&\quad \text{ such that either } i \ne j \text{ or } k \ne l, \\
&(iv). Cov\left(\varepsilon_k^i, u_l^j\right) = 0, \text{ for all } i, j, k \text{ and } l \\
&(v). Cov\left(x_{jk}^i, \varepsilon_l^j\right) = 0, \text{ for all } i, j, k \text{ and } l \\
&(vi). Cov\left(x_{jk}^i, u_l^j\right) = 0, \text{ for all } i, j, k \text{ and } l
\end{aligned}
\right\} \tag{8.5}
$$

Also, x_j^i is always independent of ε_j^i and u_j^i. Then the assumptions in (8.4) in view of (8.5) are,

$$
\left.
\begin{aligned}
E(\beta_j^i) &= \beta_j^i, \text{ for all } i \text{ and } k \\
Var(\beta_j^i) &= \sigma_k^2, \text{ for all } i \text{ and } k
\end{aligned}
\right\} \tag{8.6}
$$

Under (8.4), (8.5) and (8.6) - i.e. for the i'th cluster the variance term σ^2 in (8.2) is $(\sigma_i)^2$, the regression model (8.3) may be written as[4]

---

[3] Note that there is an identification problem if the intercept term is also assumed to follow (8.4).

[4] If we assume $\varepsilon_k^i = 0$ then the inter-cluster random coefficient model reduces to a model with fixed coefficients and heteroscedastic variances. If we further assume $\varepsilon_k^i = 0$, and $(\sigma_k^i)^2 = \sigma^2$, then it will become a special case of Bhatti's (2000) SSMC and SMC models.

$$y_j^i = \sum_{k=1}^{p} \beta_k x_{jk}^i + V_{j,\ldots i=1,2,\ldots c}^i \tag{8.7}$$

where

$$V_j^i = \sum_{k=2}^{p} \varepsilon_k x_{jk}^i + u_j^i \tag{8.8}$$

Note that (8.5) and (8.6) imply that the mean, variance and covariance's of V_j^i are as follows,

$$
\begin{aligned}
E\left(V_j^i\right) &= E\left(\sum_{k=2}^{p} \varepsilon_k^i x_{jk}^i + u_j^i\right) \\
&= E\left(\sum_{k=2}^{p} \varepsilon_k^i x_{jk}^i\right) + E\left(u_j^i\right) \\
&= 0
\end{aligned}
$$

and

$$
\begin{aligned}
E\left(V_j^i V_t^s\right) &= E\left[\left(\sum_{k=2}^{p} \varepsilon_k^i x_{jk}^i\right)\left(\sum_{k=2}^{p} \varepsilon_k^s x_{tk}^s\right)\right] + E\left[u_j^i u_t^s\right] \\
&= \delta_{is}\left(\sum_{k=2}^{p} \sigma_k^2 x_{jk}^i x_{tk}^i\right) + \left(\sigma^i\right)^2 \delta_{is}\left\{\rho + (1-\rho)\delta_{jt}\right\} \\
&= \left(\delta^1\right)^2 \delta_{is}\left\{\left(\sum_{k=2}^{p}\left(\frac{\sigma_k^2}{\left(\sigma^1\right)^2}\right) x_{jk}^i x_{tk}^i\right) + \frac{\left(\sigma^i\right)^2}{\left(\sigma^1\right)^2}\left\{\rho + (1-\rho)\delta_{jt}\right\}\right\} \\
&= \left(\delta^1\right)^2 \delta_{is}\left\{\left(\sum_{k=2}^{p} \lambda_k x_{jk}^i x_{tk}^i\right) + \mu^i\left\{\rho + (1-\rho)\delta_{jt}\right\}\right\}
\end{aligned}
\tag{8.9}
$$

where[5],

$$\lambda_k = \frac{\sigma_k^2}{\left(\sigma^1\right)^2}, \text{ for } k = 2,....p \text{ and} \mu^i = \frac{\left(\sigma^i\right)^2}{\left(\sigma^1\right)^2}, i = 1, 2, ..., c.$$

For ease of notation, the term $\left(\sigma^1\right)^2$, the disturbance variance for the first cluster, in the remainder of the section would be denoted by σ^2. Therefore, the model (8.7) can be more compactly written in matrix form as

$$y = X\beta + v \tag{8.10}$$

Now, let $\text{var}(V^i) = \sigma \Omega^i$ for the i'th cluster then, (8.9) implies that

$$\Omega^i = D^i + F^i$$

where D^i is the matrix whose (j,t)'th element is obtained by

$$\sum_{k=2}^{p} \lambda_k x_{jk}^i x_{tk}^i \tag{8.11}$$

and F^i is of the form

$$F^i = \mu^i \left[\left(1 - \rho\right) I_{m(i)} + \rho E_{m(i)} \right] \tag{8.12}$$

where $I_{m(i)}$ is the $m(i) \times m(i)$ identity matrix and $E_{m(i)}$ is the matrix whose elements are all equal to one.

---

[5] Swamy's (1970) model ignores equicorrelation within cluster and considersn heteroscedasticity i.e. $\left(\sigma_k^i\right)^2$; k=1,...,p. Whereas, in our model the variance terms of the diagonal elements of i'th cluster are $\left(\sigma^i\right)^2$ i.e. constant within clusters.

It follows from (8.11) and (8.12) that $v \ \square \ N(0, \sigma^2 \Omega)$ where $\Omega = \displaystyle\bigoplus_{i-1}^{c} \Omega^i$

is cluster diagonal. For any X matrix in (8.10), the distribution of y is

determined by $\beta, \sigma^2, \rho, \lambda = \left(\lambda_2, ..., \lambda_k \right)'$ and $\mu = (\mu^1, \mu^2, ..., \mu^c)'$

If we assume $\varepsilon_k^i = 0$, then the inter-cluster random coefficient model reduces to a model with fixed coefficients and heteroscedastic variances. If we further assume $\varepsilon_k^i = 0$, and $(\sigma^i)^2 = \sigma^2 \ (\sigma^i)^2 = \sigma^2$, then it will become a special case of SSMC and SMC models of chapters three and four, respectively.

In the next subsection we will develop an algorithm in the spirit of Ansley (1979) and King (1986) for finding the true maximum likelihood estimates of the unknown, under the constraints that λ's and μ's being strictly positive. This solves the problem of negative estimates of variances of the coefficients. However, this boundary condition of having only non-negative variances means that standard methods of estimating the standard errors of the estimates cannot be used. A solution for this is to use the bootstrap method to estimate standard errors of the estimates. An overview of the theory and application of bootstrapping is provided by Efron (1979, 1982), Efron and Tibshirani (1986), while Raj (1989), Brooks and King (1994) and Bhatti (2004) discussed its application to random coefficient models.

An Algorithm for Full Maximum Likelihood Estimation

As we know from the previous section, $v \ \square \ N(0, \sigma^2 \Omega)$ where Ω is a cluster diagonal matrix. It is important to note that the matrix Ω has the Cholesky decomposition $\Omega = LL'$, such that

$$L = \begin{pmatrix} T^1 & 0 & . & . & 0 \\ 0 & T^2 & . & . & . \\ . & . & T^3 & 0 & . \\ . & . & 0 & . & . \\ 0 & . & . & . & T^c \end{pmatrix};$$

where T^i is an $m(i) \times m(i)$ lower triangular matrix of the i'th cluster, i.e.

$$\Omega^i = T^i {T^i}'$$

If w^i_{jk} denotes the (j,k)'th element of Ω^i then non-zero elements of T^i can be found by the following recursive scheme,

$$t^i_{11} = (\omega^i_{11})^{1/2}$$

(8.13a)

$$t^i_{j1} = \omega^i_{j1} / t^i_{11}, \text{where } j = 1, \ldots, m(i.)$$

(8.13b)

The off-diagonal elements of T^i can be obtained by the formula

$$t^i_{jk} = \frac{1}{t^i_{kk}} \left(\omega^i_{jk} - \sum_{p-1}^{k-1} t^i_{kp} t^i_{jp} \right), for \ k < j, \ldots.$$

(8.13c)

whereas, the diagonal elements can be obtained by the formula

$$t^i_{jj} = \left(\omega^i_{jj} - \sum_{k=1}^{j-1} (t^i_{jk})^2 \right)^{1/2}$$

(8.13d)

Now, the model (8.10) can be transformed as

$$L^{-1}y = L^{-1}X\beta + L^{-1}V,$$

which may conveniently be written as,

$$\tilde{y} = \tilde{X}\beta + \tilde{V}$$

(8.14)

such that $\tilde{V} \ \square \ N(0, \sigma^1 I_n)$. The transformed vector $\tilde{y}$ can be obtained by the recursive relationship

$$\tilde{y}^i_l = y^i_1 / t^i_{11}$$

(8.14a)

and so on, the j'th value of the i'th cluster of y be

$$\tilde{y}^i_j = \frac{1}{t^i_{jj}}\left(y^i_j - \sum_{k=1}^{j-1} t^i_{jk}\tilde{y}^i_j\right). \tag{8.14b}$$

Similarly, the matrix X may be transformed recursively as

$$X^i_{lk} = X^i_{lk} / t^i_{11} \text{ where } k = 1, 2, ..., p \tag{8.14c}$$

and so on. The (j,k)'th element of i'th cluster of transformed matrix $\tilde{X}$ can be obtained by

$$X^i_{lk} = \frac{1}{t^i_{jj}}\left(X^i_{jk} - \sum_{l=1}^{j-1} t^i_{j1}\tilde{X}^i_{lk}\right). \tag{8.14d}$$

The log likelihood of the original model (8.10) and the transformed model (8.14) is

$$\begin{aligned}\ell\left(\beta,\sigma^2,\rho,\lambda,\mu\right) &= const - \frac{n}{2}\log\sigma^2 - \frac{1}{2}\log\left(\Omega\right(\\ &\quad -\frac{1}{2\sigma^2}\left(y - X\beta\right)'\Omega^{-1}\left(y - X\beta\right)\\ &= const - \frac{n}{2}\log\sigma^2 - \log\left(L\right(\\ &\quad -\frac{1}{2\sigma^2}\left(\tilde{y} - \tilde{X}\beta\right)'\left(\tilde{y} - \tilde{X}\beta\right),\end{aligned} \tag{8.15}$$

where the determinant of the i'th cluster of L can easily be calculated by

$$\left.\begin{aligned}|L| &= \prod_{i=1}^{c}\left(|T|\right)\\ \text{and}\\ |T^i| &= \prod_{j=1}^{m(i)} t^i_{jj}\end{aligned}\right\} \tag{8.16}$$

Setting $\partial \ell(\beta, \sigma^2, p, \lambda, \mu) / \partial \beta$ and $\partial \ell(\beta, \sigma^2, p, \lambda, \mu) / \partial \sigma^2$ to zero, we obtain

$$\hat{\beta} = (\tilde{X}'\tilde{X})^{-1}\tilde{X}'\tilde{y} \tag{8.17}$$

$$\hat{\sigma}^2 = \hat{\eta}'\hat{\eta} / n \tag{8.18}$$

respectively, where

$$\hat{\eta} = (\hat{y} - \tilde{X}\hat{\beta}) \tag{8.19}$$

is the OLS residual vector from (8.14). Substituting (8.17) and (8.18) into (8.15) yields the concentrated log likelihood

$$\begin{aligned}
\tilde{\ell} &= c - \frac{n}{2}\log \hat{\eta}'\hat{\eta} - \log|L| \\
&= c - \frac{n}{2}\left[\log\left\{\hat{\eta}(|L|)\right\}^{1/n}\right] \\
&= c - \frac{n}{2}\left[\log \hat{e}'\hat{e}\right] \\
&= c - \frac{n}{2}\log \sum_{j=1}^{n}(\hat{e}_j)^2
\end{aligned}$$

where c is a constant and

$$\hat{e}_i = \hat{\eta}(|L|)^{1/n}. \tag{8.20}$$

Therefore, the estimation problem reduces to minimizing the sum of squares of

$$\hat{S} = \sum_{i=1}^{n}\hat{e}i^2$$

with respect to ρ, λ_k ; ($k=2,...,p$) and μ^t; ($i=2,...,c$), given that $\mu^1=1$. This may be obtained by using a standard nonlinear least squares algorithm to minimize $\hat{S}$, where for any given ρ, λ_k and μ^i, $\hat{e}_i$ ($i=1,2,...,n$) are obtained as follows:

i.Transform y and X to $\hat{y}$ and $\tilde{X}$, respectively, using (8.10), (13a-13d), (8.14), (14a-14d) and at the same time progressively calculating $|L|$ via (8.16)

ii.Compute $\hat{\beta}$ and $\hat{\eta}$ using (8.17) and (8.19) and the use of (8.20) to calculate $\hat{e}_i$.

When the values of $\hat{p}$, $\hat{\lambda}$ and μ, which minimizes $\hat{S}$ have been found, the MLE of β is $\hat{\beta}$ from step (ii) of the final iteration, while σ^2 can be evaluated by using the final value of $\hat{\eta}$ in (8.18). This procedure converges to a solution of the first-order conditions for maximizing (8.15), that may or may not correspond to the global maximum. Oberhofer and Kmenta (1974) provide further discussion of this.

8.3. APPLICATIONS OF THE STOCHASTIC MODEL

In this section we use the Cobb-Douglas type production function to estimate unknown stochastic parameters, by using the random coefficients model discussed in the previous subsections. The random coefficients model is applied to Bangladeshi agricultural data for the year 1996-97, collected from 600 farms of the seven selected districts of Khulna and Rajshahi divisions. For details of the data refer to table 8.2. Due to computational ease and brevity, we will consider two independent random samples of the size of 20 observations (farms) from each district of both divisions. The detailed sampling design and measurement of variables are given in table 8.2.

The model (8.7), in the form of unrestricted Cobb-Douglas production functions (linear in the logarithms), under (8.4), (8.5) and (8.6), and can generally be written as

$$y_j^i = \sum_{k=1}^{p} \beta_k x_{jk}^i + V_{j,\dots}^i \qquad (8.21)$$

where y_j^i is the log of output for the j'th farm in the i'th cluster, and x_{jk}^i is the log of a (1xp) vector of functions of values of inputs associated with the j'th farm on the i'th district, such that the first element of the x_{jk}^i is assumed to be one. Thus, we can write model (8.21) as

$$\log(y_j^i) = \beta_1 + \beta_2 \log (X_{j2}^i) + \beta_3 \log (X_{j3}^i)\beta_4 \log (X_{j4}^i) + V_j^i, \quad (8.22)$$

where

y_j^i : Total value of output per acre for j'th farm in the i'th cluster, in thousands Taka,

x_{j2}^i : Labour input per acre in man-days of j'th farm in the i'th cluster,

x_{j3}^i : BC (biological-chemical) input per acre in thousands Taka, of j'th farm in i'th cluster,

x_{j4}^i : Size of j'th farm in i'th cluster (district) in acres,

and the random variable y_j^i has the distributional properties as in (8.9), (8.10), and (8.11).

Using the likelihood and the concentrated likelihood functions (8.15) and (8.20), respectively, the Cobb-Douglas type production function (8.21), and (8.22) particularly, we can obtain the estimates of the unknown parameters by Levenberg - Marguardt algorithm. Thus, the estimated model (8.22) based on the first set of 140 randomly selected farms is

$$\log\left(y_j^i\right) = -0.272 + 0.675\log\left(X_{j2}^i\right) + 0.0351\log\left(X_{j3}^i\right) + 0.0347\log\left(X_{j4}^i\right) .$$
$$(8.23)$$

Similarly, the estimated model (8.22) based on the second set of 140 selected farms is

$$\log\left(y^i_j\right) = -0.269 + 0.674\log\left(X^i_{j2}\right) + 0.033\log\left(X^i_{j3}\right) + 0.04\log\left(X^i_{j4}\right)$$

$$(8.24)$$

The estimated values of the parameters $\mu^i = \dfrac{(\sigma^i)^2}{\sigma^2}$, for $i=2,...,7$ clusters from both the samples are given in table 8.1 below.

Table 8.1. Estimated Values of the Parameters Based on Models (8.23) and (8.24)

Parameters	$\hat{\sigma}$	$\hat{p}$	μ^2	μ^3	μ^4	μ^5	μ^6	μ^7
Sample 1 from Model (8.23)	0.051	0.0017	0.381	0.223	0.024	0.049	0.078	0.082
Sample 2 from Model (8.24)	0.0472	0.0065	0.363	0.22	0.028	0.043	0.083	0.082

The estimated results in table 8.1 demonstrate that there exists a high district-wise heteroscedasticity in all the seven districts of Khulna and Rajshahi divisions. This is due to varying geographical, cultural and economic factors, with their also being differences in the farmers' motivations in the two divisions. The Khulna division is the second largest industrial division of the country and therefore, one can expect extra farm job openings, which would make the opportunity cost of labour much higher compared to that in the Rajshahi division.

The main reason for poor economic results in all districts of Rajshahi division is due to its geographical location, as the division is wholly dependent on agricultural based industries, which are always affected by climate and weather conditions, e.g. cyclones, floods and droughts. During the monsoon, an excess of water in the Rajshahi division affects its division and causes considerable damage to its cultivated lands. This restricts farmers to grow particular crops. Moreover, during the dry season, this division faces a shortage of water (see, Hossain (1990)). This creates a problem with irrigation and, hence, watering the plants involves great capital investment in the form of deep tube-wells. Farming activities in the Rajshahi are therefore different from that of Khulna division, and are hence the result of high cluster-wise heteroscedasticity. The estimates of the parameters λ_k's which are the ratio of the variances of random coefficients to the variance of the error term of the first cluster, i.e. $\lambda_k = \sigma_k^2 / \sigma^2$ for $k=2,...,4$, are given in table 8.2 below (see equation (8.9) for the reparameterisation of the λ's and μ's in our model).

Table 8.2. Some Estimates of Random Coefficient Model

K	β_k Sample 1	β_k Sample 2	λ_k Sample 1	λ_k Sample 2	σ_k Sample 1	σ_k Sample 2
2	0.683	0.681	0.2194	0.2076	0.0191	0.017
3	0.042	0.042	0.0622	0.0155	0.0112	0.0101
4	0.043	0.048	0.5217	0.4710	0.0233	0.0311

The estimates given in model (8.23), (8.24) and in tables 8.1 and 8.2 provide a local maximum for the criterion function, but they do not necessarily correspond to the global maximum required.

The high elasticity of labour as compared to other variables in models (8.23) and (8.24) is due to the following reasons. Firstly, Bangladesh is a labour intensive country and labour is very cheap compared to land and/or BC inputs, etc. Secondly, the shape and small size of the farms prevent owners from utilizing new technology in order to gain high output levels. In such environments labour contributions could of course be more important than capital (i.e. land and BC inputs). However, one should not forget that there must be a limit for the number of labourers working for a given size of the farm. On the other hand, the estimated coefficients of BC inputs and for that of farm size are 0.043 and 0.042 in model (8.23), and 0.042 and 0.048 in model (8.24), respectively. These results support the empirical work reported by Hoque (1991), Wu and Bhatti (1994), Parikh (2000), Parikh and Radhakrishna (2002) and Ali (2002).

The 2SLR Model

The model (8.1), under (8.2) is similar to that of the random effects model or one-way error component model used by econometricians and other social scientists in the analysis of panel and/or time-series data.

A simple (re) formulation appropriate in this case is given in equation (5.1)

$$y_{it} = \sum_{k=1}^{p} \beta_k x_{itk} + u_{it},$$
$$(i = 1, 2, ..., N, t = 1, 2, ..., T)$$

where

$$u_{it} = \mu_i + v_{it}, \tag{8.25}$$

in which $i=1,2,...,N$, where N stands for the number of individuals (e.g. households) in the sample and $t=1,2,...,T$, where T stands for the length of the observed time series. Each of the $\mu_i's$, ($i=1,...,N$) are called an individual effect and v_{it} is the usual (white noise) error term. In this reformulation (at this stage) it is assumed that every cluster has the same number of observations (T). Note that this structure is similar to that of (2.9) with $T = c$, clusters under the assumptions (5.2).

The pioneers Balestra and Nerlove (1966), Wallace and Hussain (1969) and Maddala (1971) have drafted the basic outline of the model (8.4). They assume that:

1   The random variables μ_i and v_{it} are mutually independent.

2   2.$E(u_{it}) = 0$. This implies that $E(\mu_i) = 0$ and $E(v_{it}) = 0$

3   $\text{var}(\mu_i) = \begin{cases} \sigma_\mu^2, \text{ for } i = i' \\ 0, \text{ otherwise} \end{cases}$

$\text{var}(v_{it}) = \begin{cases} \sigma_v^2, \\ for\ i = i', t = t' \\ 0, \text{ otherwise.} \end{cases}$

4

In comparing (8.1) with (8.25), it is noted that $u_{it} = \mu_i + v_{it}$, $\sigma^2 = \sigma_\mu^2 + \sigma_v^2$, and $\rho = \sigma_\mu^2 / \sigma^2$. The only difference between (5.2), (8.1) and (8.25) is that in model (8.25) the ith cluster consists of the time-series of the ith individual and the number of observations in a 'cluster' is T, the length of the time series. In the econometrics literature, this model is also called the one-way error component model. This model is frequently used to model panel data in the econometrics literature as noted in chapter five, and is referred to as 2SLR model. Empirical comparison of estimates of fixed (8.25) and random coefficient models (8.23) and (8.24) are given in table 8.3 below.

Table 8.3. Some Estimates of Random and Fixed Coefficient Models

K	β_k (8.23) (8.24) (8.25)	Standard Error, σ_k (8.23) (8.24) (8.25)	T (3.25)
2	0.683 0.681 0.564	0.021 0.020 0.112	5.0748
3	0.042 0.042 -0.044	0.011 0.005 0.069	-1.0289
4	0.043 0.048 0.0867	0.032 0.030 0.053	1.04778

Surprisingly, fixed coefficient model (8.25) reconfirm our results of the stochastic fitted models (8.23) and (8.24) for the high elasticity of labour contribution. However, negative relationship between farm size outputs may explains the stochastic patterned in the data structure and labour redundancy theory with increase in technology and the quality of BC, for example a typical case of underdeveloped countries like Bangladesh.

8.4. SUMMARY OF TESTS FOR CLUSTER EFFECTS

It is important to note that if we assume $\varepsilon_k^i = 0$ then the inter-cluster random coefficient model (8.7) will reduce to our 2SLR model, which is discussed in (8.2) and (8.25). For the sake of simplicity, we can express (8.25) in matrix form as follows,

$$y(i) = X(i)\beta + u(i);$$

(8.26)

for which $i = 1,...,c = m$, (number of clusters), and where $y(i)$ is an $m(i) \times 1$ vector of y_{ij} values, $X(i)$ is an $m(i) \times p$ matrix whose (j,k)th element is x_{ijk} and $u(i)$ is an $m(i) \times 1$ vector of u_{ij} values. If y and u denote the $n \times 1$ vectors of stacked $y(i)$ and $u(i)$ vectors respectively, and X is the $n \times p$ stacked matrix of $X(i)$ matrices, then (8.26) can be written as

$$y = X\beta + u$$

(8.27)

such that

$$u \sim N\left(0, \sigma^2 \Delta(\rho)\right),$$

where $\Delta(\rho) = \overset{m}{\underset{i=1}{\oplus}} \Delta_i(\rho)$ is cluster diagonal, with sub matrices

$$\Delta_i(\rho) = (1-\rho) I_{m(i)} + \rho E_{m(i)} \qquad (8.28)$$

in which $I_{m(i)}$ is an $m(i) \times m(i)$ identity matrix and $E_{m(i)}$ is an $m(i) \times m(i)$ matrix with all elements equal to one. The structure of the above models (8.26) to (8.28) are similar to the 2SLR models in chapter five. Moreover, it is important to note that if we assume in (8.25) or (8.26) that the error term $\varepsilon_k^i = 0$, and the variances for each cluster are the same i.e., $(\sigma^i)^2 = \sigma^2$, then the 2SLR model becomes a special case of Bhatti's (2000) SSMC and SMC models, as has been discussed in earlier chapters. Therefore using the invariant approach, $y \rightarrow \gamma_0 y + X\gamma$, and the maximal invariant vector, υ, E can reach to the following theoretical propositions.

Proposition 8.1.

The $s(\rho_1)$ test is approximately UMPI for all values of p and m. It is worth noting that the test, $s*$ for testing simple null against simple alternative can be expressed in terms of the standard F distribution, as follows

$$\Pr\left[s* < c_\alpha(1-\rho_1)\right] = \Pr\left[\frac{1}{s*} > \frac{1}{c_\alpha^*(1-\rho_1)}\right]$$

$$= \Pr\left[\frac{\xi_1^2}{\sum_{i=2}^{n-p}\xi_i^2} < \left(\frac{1}{c_\alpha^*(1-\rho_1)} - 1\right)\right]$$

$$= \Pr\left[\frac{\chi_{(1)}^2}{\chi_{(n-p-1)}^2 \big/ (n-p-1)} > \left(\frac{1}{c_\alpha^*(1-\rho_1)} - 1\right)(n-p-1)\right]$$

$$= \Pr\left[F(\vartheta_1, \vartheta_2) > F_\alpha\right],$$

where

$$F_\alpha = \left(\frac{1}{c_\alpha^*(1-\rho_1)} - 1\right)(n-p-1), \quad \vartheta_1 = 1, \vartheta_2 = n-p-1$$

and $F(\vartheta_1, \vartheta_2)$ denotes the F distribution with ϑ_1 and ϑ_2 degrees of freedom. If F_α is the $100(1-\alpha)$ percentile of the F distribution (obtainable from standard tables), then the approximate critical value, c_α^* can be obtained by the following equation,

$$c_\alpha^* = \frac{1}{(1-\rho_1)\left(\dfrac{F_\alpha}{(n-p-1)} + 1\right)}.$$

Proposition 8.2.

If Y follows two-stage SSMC model i.e.,

$$Y \sim N\big(0, \Delta(\rho)\big),$$

where covariance matrix is given in (8.27), then testing $H_0 : \rho = 0$ against $H_a : \rho > 0$ will reject null for the small value of $r(\rho_1) = Y'\big(\Delta^{-1}(\rho_1) - I_n\big)Y < c_\alpha$, which in chapter three is called as BO test for the two-stage SSMC model which is given by $r(\rho_1)$ which is also LMP and LB test.

Proposition 8.3.

In case where cluster diagonal elements are different in (8.27) for each cluster, then the testing problem will be of testing, $H_0 : \rho_1 = \rho_2 = ... = \rho_m = 0$, against, $H_a : \rho_1 \geq 0, \rho_2 \geq 0, ..., \rho_m \geq 0$,

which will reject the null for small values of

$$d = Y'AY = Y'(I_n - D)Y.$$

This is the LMMP test, and is equivalent to the LB test as well.

Proposition 8.4.

If Y follows a two-stage SMC model, i.e., $Y \sim N(0, \sigma^2 \Delta(\rho))$, then testing $H_0 : \rho = 0$ against $H_a : \rho > 0$ is to reject null for small values of

$$s(\rho_1) = \frac{Y'\Delta^{-1}(\rho_1)Y}{Y'Y} < c_\alpha$$

The $s(\rho_1)$ test is the BOI, with UMPI, and multistage and multiparametric LMMPI tests also be derived on a similar way as presented in in chapters 3 and 4.

8.5. CONCLUDING REMARKS

In this chapter we have developed and illustrated efficient estimation procedures for the LR model with fixed and stochastic coefficients, cluster-wise heteroscedasticity and intra cluster correlation, based on two-stage-clustered survey data. We applied a two-stage clustered sampling procedure, restricted by computational convenience, to Bangladesh data from the selected districts of the Khulna and Rajshahi divisions. This chapter links between previous chapters which focused on 2SLR, 3SLR and LNSG models and summarised propositions associated with BO, BOI, PO, POI and LMMPI tests for the hypothesis that the intra cluster, ρ, has any given value.

9. TESTING FOR CHANGE POINT

9.1. INTRODUCTION

Change point problems arise in a variety of experimental and mathematical sciences, as well as in economics, engineering and health sciences. Such errors may also occur in the area of statistical quality control, where when one observes the output of a production line there would be a wish to detect and signal deviation from the acceptable level, while observing data. For example, in the field of epidemiology, one may be interested in testing whether the incidence of a disease has remained constant over time, and if not, attempt to estimate the times of changes in order to suggest possible causes. In the area of Electro-cardiogram whereas, one may be interested in detecting the change point to understand rhythm and pattern recognition, whilst in meteorology, one may be interested in detecting possible changes in climate or in weather conditions. Detection of change points can also be of interest in history and in signal processing, economics, business management, and for industries, such as for example the exploration and detecting the depth of an oil field.

Historically the issue of change point problems began with Maguire et al's (1952) article, in which they considered 109-time intervals between explosions in British Coal mines between 1875 to 1950, which killed more than 10 people. Maguire et al concluded in their study that the mines explosion data follow a one-parameter exponential distribution with constant scale parameter over time, and that the change point in the scale parameter was after 1895. Some years after Maguire et al's study, the change point problem became an established research area, with series of articles published by Page (1954, 1955 and 1957) who considered the change point problem in the context of quality control, and in determining the quality of the output from a continuous production process. Page's non-parametric tests for change point are based on cumulative sums (CUSUM) of square method.

Page's work was all that appeared on change point problems during the 1950's. A decade after her first published work, there was a burst in research and interest in the area, based on mostly parametric statistical procedures. For example, Chernoff and Zacks (1964) used the Bayesian inference to derive tests for testing whether a change has occurred in the mean of normally distributed independent random variables. The power comparison of Bayesian and Page's (1955) sign tests by Chernoff and Zacks in their study demonstrated the superiority of the Bayesian test in cases other than if the change had occurred near the very beginning of the sample. The results of Chernoff and Zacks' study was extended by many authors like Kander and Zacks (1966),and Hinkly (1970), Hsu (1979), Worsley (1983, 1986), Levin and Kline (1985), Andrews (1993), Luo et al. (1997), Belisle et al. (1998) and Bhatti and Wang (1998) into different directions.

Hinkley and Hinkley (1970) considered the problem of testing for a change in mean of binomial proportion. Hsu (1979) whereas looked into the problem of a change in mean of a sequence of gamma random variables, with applications to the variation of stock market returns, whilst Levin and Kline (1985) considered a CUSUM test for epidemic change in spontaneous abortion epidemiology. Luo et al. (1997) developed likelihood ratio test for a change point in survival data, subject to right censoring. Belisle et al. (1998) whereas introduced a change point test for integrated valued time series models. Among the other notable researchers who contributed to the area of change point are Maguire et al (1952), Hinkley (1970), Scott and Knott (1974), Hawkins (1977), Worsley (1983, 1986), Cox and Spjotvoll (1982) and Schechtman (1983). An extensive bibliography of this subject is given by Csorgo and Horvath (1988), Krishnaiah and Mio (1988), Huskova and Sen (1989), Zacks (1983, 1991), and Antoch and Huskova (1992), among others.

The main objective of this chapter is to consider tests for testing change point problems of variances in the sequence of independent random variables with unknown means. This chapter conducts an empirical power study of the five tests by noting that the underlying testing problem is a member of the wider class of tests in which the nuisance parameter (the vector of change point) disappears under the null hypothesis. There is a growing literature on this class of testing problem, with a few examples of important contributions on this topic including Andrews et al. (1993), Bhatti (1995b), Bhatti and Wang (1998) and more recently Bhatti (2000).

This chapter derives the distribution of the B-test and conducts a power study of the L, B, R, LM and C-tests to see which has the highest power in order to be called an approximately uniformly most powerful test for this testing problem. Among the tests looked in this study, the first is based on the Lehmann U-statistic (see, for details, Lehmann (1951)). The second test is derived using the Bayesian method, which was initially proposed by Kander and Zacks (1966) and later followed up by Hsu (1979) for testing the change point problem of the variance in a sequence of normal random variables with a known mean. The third, R-test, whereas is constructed using the maximum likelihood method. The likelihood procedure for the change point problem was initially applied during the late 1950's to late 1970's by Hinkley (1969, 1970, 1971) and Hawkins (1977) among others.

The fourth, C-test look at in this study is based on CUSUM of square tests. Basics of this test are given by Page in her earlier work. Hinkly (1971) noticed that the CUSUM estimate is asymptotically biased and hence is less efficient than the maximum likelihood estimate (MLE). Brown et al's. (1975) first proposed the CUSUM of square test as a test for regression coefficients, which was then followed by Hansen (1991) and Ploberger (1989) who showed that the CUSUM of square test is essentially a test for the change in variance. The fifth and final test is the LM-test, which was first proposed by Nyblom (1989) to test for the constancy of regression coefficients and was generalized by Hansen (1991) to test for the change in variance. Nybolm (1989) and Hansen (1991) provide a procedure of finding critical values of the LM test.

For this chapter we will only concentrate in computing the power of these five tests for our study. The rest of this chapter is as follows. In the next section the testing problem and the test statistics, namely, L, B, R, C and LM, some selected approximate critical values are calculated using simulation distribution under the null hypothesis. In the third section, we will calculate and compare the power of these tests. Our findings from this calculation and comparison suggest that the power gain among these tests is not clear overall parametric space. However, it is observed that the B- and the CUSUM tests are marginally superior to those of the L, LM, R-tests.

9.2. THE TESTING PROBLEM AND THE TESTS

Suppose $X_1, ..., X_n$ is a sequence of independent random variables, such that the first τ of this sequence, $X_1, ..., X_\tau$ has a common normal distribution[1] $N(\mu, \sigma_1^2)$, the last n of this sequence, $X_{\tau+1}, ..., X_n$ has a common normal distribution $N(\mu, \sigma_1^2)$, where the mean μ and standard deviations σ_1, σ_2 are all unknown, and the integer τ, which is called the change point, is also unknown. Throughout this chapter our interest is in testing the null hypothesis,

$$H_0 = \sigma_1 = \sigma_2 \text{ (i.e., no change)} \tag{9.1}$$

against

$$H_1 = \sigma_1 \neq \sigma_2 (\text{or } \sigma_1 < \sigma_2, \text{ or } \sigma_1 > \sigma_2). \tag{9.2}$$

To test the above problem, we need to construct the L, B, R, C and the LM test statistics in the following subsections.

The Lehmann U-Statistics

This subsection consider the problem of testing (9.1) against (9.2), which is to reject (9.1) for large values of the test statistic L, such that

$$L = \max |L_k|$$
$$2 \leq k \leq n-2$$

where L_k is obtained by the expression,

$$L_k = \frac{U_{k,n-k} - 1/2}{\sqrt{D(U_{k,n-k})}} \tag{9.3}$$

---

[1] These normal data may arise from SMC and SSMC normal distributions.

In (9.3) above $U_{k,n-k}$ is the Lehmann's non-parametric two-sample U-statistic (see Lehmann (1951)) of $X_1,...,X_k$ and $X_{k+1},...,X_n$. This is created with the symmetric kernel

$$h(X_1,X_2;X_{k+1},X_{k+2}) = I_{\left(\left|X_{k+2}-X_{k+1}\right|>\left|X_2-X_1\right|\right)}$$

by forming

$$U_{k,n-k} = \left\{\binom{k}{2}.\binom{n-k}{2}\right\}^{-1} \cdot \sum_{\substack{1\leq i<j\leq k \\ 1\leq 1<m\leq n-k}} h(X_i,X_j;X_{k+1},X_{k+m})$$

$$(9.4)$$

The denominator of L_k in (9.3), namely, $D(U_{k,n-k})$ is the variance of the $U_{k,n-k}$ structure given by (9.4) when H_0 is true (i.e., $1=\sigma_2$). Note that when H_0 is true the mean of (9.4) can be easily show as

$$E(U_{k,n-k}) = 1/2.$$

This can be proved with the help of the following lemma (9.1) given below.

Lemma 9.1.
If $U_{k,n-k}$ is defined by (9.4), then

$$D(U_{k,n-k}) = \left\{\binom{k}{2}\binom{n-k}{2}\right\}^{-1} \left\{\left[(k-2)(k-3)+(n-k-2)(n-k-3)\right](0.0417)\right.$$
$$+(k-2)(n-k-2)(n-6)(0.0138)+(k-2)(n-k-2)(0.0836)$$
$$\left.+(n-4)0.2008+0.25\right\}$$

when H_0 is true (i.e., $\sigma_1=\sigma_2$). (The detailed proof of the above proposition is given by Bhatti and Wang (1998) and summarised in appendix 9B at the end of this chapter for ready reference of the readers).

For testing H_0: $\sigma_1 = \sigma_2$ against $H_1 : \sigma_1 < \sigma_2$, with unknown change point, k, we reject H_0 for large values of LL, where

$$LL = \max_{2 \le k \le n-2} L_k$$

Similarly, for testing H_0: $\sigma_1 = \sigma_2$ against $H_1 : \sigma_1 > \sigma_2$, with unknown change point, k, we reject H_0 for small values of $\overline{LL}$, where

$$\overline{LL} = \min_{2 \le k \le n-2} L_k$$

Occasionally, to avoid misunderstanding, we shall use

$$L_k(X_1, ..., X_n), LL(X_1, ..., X_n), \overline{LL(X_1, ..., X_n)},$$

which are calculated based on the sequence $X_1, ..., X_n$. Let $Y_i = X_{n+1-i}, i = 1, ..., n$.

Therefore,

$$L_k(X_1, ..., X_n) = -L_{(n-k)}(Y_1, ..., Y_n)$$

which implies that

$$LL(X_1, ..., X_n) = -\overline{LL}(Y_1, ..., Y_n).$$

As the null distribution of $X_1, ..., X_n$ and $Y_1, ..., Y_n$ are identical, then the null distributions of LL and $-\overline{LL}$ are also identical. Therefore, we only need to obtain the critical values of LL, which can be computed by the simulation experiment, briefly detailed below.

In this the simulation experiment we generated 5000 samples of various n values and simulated the null distributions of L and LL tests in order to obtain approximate critical values. The required critical values of L and LL tests with level of significance denoted by L, and LL, respectively, and their computed values tabulated in table 9.1 of appendix 9A.

Bayesian Test Statistics for Testing Change Point

When the mean is known, we can let it equal zero without loss of generality. On the change point problem of the variance, Hsu (1979) proposes the test statistic T, defined in (9.5) below

$$T = \sum_{i=1}^{n} \frac{(i-1) X_i^2}{(n-1) \sum_{i=1}^{n} X_i^2} .$$

$$(9.5)$$

This test is derived from the Bayesian test statistic proposed by Kander and Zacks (1966) under the assumption that the initial level of the standard deviation, σ_1, is known. When the mean is unknown, it is usually estimated under H_0 by the sample average, $\overline{X}$, so we can find the test statistic to be

$$B = \sum_{i=1}^{n} \frac{(i-1)\left(X_i - \overline{X}\right)^2}{(n-1) \sum_{i=1}^{n} \left(X_i - \overline{X}\right)^2} ,$$

such that, $0 \leq B \leq 1$. Analogous to L-test, we may write B as $B\left(X_1, \cdots, X_n\right)$, and therefore let $Y_i = X_{n+1-i,}$ $i = 1, \cdots, n$. We can therefore state that,

$$B\left(X_1, \cdots, X_n\right) + B\left(Y_1, \cdots, Y_n\right) = 1 .$$

Because the null distributions of $X_1, \cdots, X_n$ and $Y_1, \cdots, Y_n$ are identical, thus B is symmetrically distributed about the mean 0.5 under H_0 .

Let

$$S = \sum_{i=1}^{n}\left(X_i - \bar{X}\right)^2, \quad V = \sum_{i=1}^{n}(i-1).\left(X_i - \bar{X}\right)^2.$$

Under H_0, without loss of generality, let $\mu = 0$, $\sigma = 1$. Then $S \sim \chi^2(n-1)$. Therefore, $V = X'ACAX$, where $X = (X_1, \cdots, X_n)$, $C = diag(0, 1, \cdots, n-1)$,

$$A = I - \frac{11'}{n}, \quad 1' = (1, \cdots, 1).$$

Because $A.1 = 0$, therefore;

$$ACA = U'\Lambda U$$

where $\Lambda = diag(\lambda_1, \cdots, \lambda_{n-1})$, λ_i, $i = 1, \cdots, n-1$, are the positive characteristic roots of ACA, U is row orthogonal matrix, i.e. $U.U' = I_{n-1}, U'$ and 1 is the orthogonal, i.e. $U.1 = 0$. Let $(Y_1, \cdots, Y_{n-1}) = UX$. It can be said that

$$V = \sum_{i=1}^{n-1} \lambda_i.Y_i^2, \quad S = \sum_{i=1}^{n-1} Y_i^2, \tag{9.6}$$

and $Y_i, \cdots, Y_{n-1}$ are mutually independent and identically distributed random variables with distribution $\chi^2_{(1)}$. Similar to T in (9.5), B is a linear function of Dilichlet variables (cf. Johnson and Kotz 1972, Ch. 40, Sec. 5).

Lemma 9.2.
The variance for the null distribution of B is

$$D(B) = \frac{n-2}{6.(n-1)^2},$$

and the standard statistic of B,

$$B^* = -2.4 \frac{B - 0.5}{\frac{1}{n-1}\sqrt{\frac{n-2}{6}}}$$

which has a negative Kourtosis measure of the form,

$$\gamma_2(B^*) = -2.4 \frac{n^3 + 6.n^2 - 19n - 6}{(n-2)^2(n+3)(n+5)}.$$

Proof: Note that under $H_0, X_1, \cdots, X_n$ are independently, normally distributed with mean μ and variance σ^2. S is sufficient and boundedly complete about μ and σ^2 and the distribution of B does not depend on μ and σ^2. According to the property of bounded completeness (see Lehmann, 1986, Ch. 5, Sec. 1, Th. 2), S is independent of B. Then

$$E\!\left(B^k\right) = \frac{E(V^k)}{(n-1)^k E(S^k)}.$$

Without loss of generality, let $\mu = 0$, $\sigma = 1$. Then $S \sim \chi^2(n-1)$ and V is defined by (9.6), i.e., V is a linear function of mutually independent and identically distributed random variables $Y_1, \cdots, Y_{n-1}$ with distribution $\chi^2_{(1)}$. We get

$$E(V^2) = \left(\sum_{i=1}^{n-1} \lambda_i\right)^2 + 2\sum_{i=1}^{n-1} \lambda_i^2$$

$$= \frac{(n-1)(n+1)(3n^2 - 4n - 1)}{12}.$$

Hence

$$D(B) = \frac{E(V^2)}{(n-1)^2 E(S^2)} - 0.5^2 = \frac{n-2}{6.(n-1)^2}.$$

Because B is symmetrically distributed about the mean 0.5, $E(B - 0.5)^3 = 0$. Hence

$$E(B^3) = \frac{3}{2}.E(B^2) - \frac{3}{4}.E(B) + \frac{1}{8} = \frac{n^2 - 3}{8(n-1)^2}.$$

We get

$$E(V^4)$$

$$= \left(\sum_{i=1}^{n} \lambda_i\right)^4 + 12\left(\sum_{i=1}^{n} \lambda_i^2\right)\left(\sum_{i=1}^{n} \lambda_i\right)^2 + 32\left(\sum_{i=1}^{n} \lambda_i^3\right)\left(\sum_{i=1}^{n} \lambda_i\right) + 12\left(\sum_{i=1}^{n} \lambda_i^2\right)^2 + 48\left(\sum_{i=1}^{n} \lambda_i^4\right)$$

$$= \frac{(n-1)(n+1)\left(15n^6 + 120n^5 + 95n^4 - 896n^3 - 791n^2 + 2504n - 471\right)}{240}$$

and

$$E\left(B - \frac{1}{2}\right)^4 = \frac{E(V^4)}{(n-1)^4.E(S^4)} - 2E(B^3) + \frac{3}{2}E(B^3) - \frac{1}{2}E(B) + \frac{1}{16}.$$

Hence

$$\gamma_2(B^*) = \frac{E\left(B - \frac{1}{2}\right)^4}{\left[D(B)\right]^2} - 3$$

$$= -2.4.\frac{n^3 + 6.n^2 - 19.n - 6}{(n-2)^2.(n+3).(n+5)}.$$

which completes the proof.

Similar to Hsu (1979), using an Edgeworth expansion (see Johnson and Kotz (1972), chapter 12, sec. 4.2), we obtain the approximate critical values for β^* under H_{0}. The percentage value of β^* is denoted by B_{p}, with these values tabulated for various values of n and α in table 9.2 appendix 9A.

For testing the null hypothesis H_0: $\sigma_1 = \sigma_2$ against the two-sided alternative, H_1: $\sigma_1 \neq \sigma_2$, we reject H_0 for large values of $\left|B^*\right|$, the critical value with level $\iota\sigma$ $B_{1-/2}$. For testing H_0: $\sigma_1=\sigma_2$ against H_1: $\sigma_1<\sigma_2$, we reject H_0 for large values of B, the critical value with level $\iota\sigma B_{1-}$. For testing H_0: $\sigma_1=\sigma_2$ against H_1: $\sigma_1>\sigma_2$, we reject H_0 for small values of B, the critical value with level $\iota\sigma B_{1-}$ are shown in table 2 of appendix 9A.

The Likelihood Ratio Test

When the mean is known, we can let it equal zero without any loss of generality. On the change point problem of the variance, for the fixed the change point

$$\tilde{R}_{\tau} = \frac{\tilde{S}_{\tau}^{\tau} \cdot (S_{\tau}^*)^{n-\tau}}{\tilde{S}^n}$$

where

$$\tilde{S}^2 = \frac{\sum_{i=1}^{n} X_i^2}{n}, S_{\tau}^2 = \frac{\sum_{i=1}^{\tau} X_i^2}{\tau}, and \left(\tilde{S}_{\tau}^*\right)^2 = \frac{\sum_{i=\tau+1}^{n} X_i^2}{n-\tau}.$$

When the mean is unknown, similarly to the B-test, we find the test statistic to be

$$\tilde{R}_{\tau} = \frac{\tilde{S}_{\tau}^{\tau} \cdot (S_{\tau}^*)^{n-\tau}}{\tilde{S}^n}$$

where

$$S^2 = \sum_{i=1}^{n} \frac{(X_i - \overline{X})^2}{n}$$

$$S_\tau^2 = \sum_{i=1}^{\tau} \frac{(X_i - \overline{X})^2}{n}$$

$$(S_\tau^*)^2 = \sum_{i=\tau+1}^{n} \frac{(X_i - \overline{X})^2}{n - \tau}$$

For testing H_0: $\sigma_1 = \sigma_2$ against H_1:$\sigma_1 \neq \sigma_2$, we reject H_0 for large values of R, where is defined as,

$$R = \max_{1 \leq k \leq n-1} R_k$$

In our simulation experiment we generated 5000 samples of size n and simulated the null distributions of R-test in order to obtain approximate critical values. The critical values of R with level α are denoted by $R\alpha$ and they are computed for various values of alpha and sample sizes. The resulted critical values are tabulated in table 9.3 of appendix 9A.

The CUSUM of Square Test

The CUSUM of squares test was first proposed by Brown et al. (1975) as a test of regression coefficients, but the results of Hensen (1991) and Ploberger (1991) show that the CUSUM of square test is essentially a test for the change in variances, as described earlier on in this chapter. In the case of a sequence of normal random variables, the test statistic can be written as

$$C = \max_{k} \left| S_k - \frac{k}{n - 1} \right|,$$

where

$$S_k = \frac{\sum_{i=1}^{k} W_i^2}{\sum_{i=1}^{n-1} W_i^2},$$

$$W_i^2 = \frac{(X_{i+1} - \overline{X}_i)^2}{\frac{i+1}{i}},$$

and

$$\overline{X}_i = \frac{\sum_{j=1}^{i} X_j}{i}.$$

For testing H_0: $\sigma_1 = \sigma_2$ against H_1 : $\sigma_1 \neq \sigma_2$, we reject H_0 for large values of C. The critical values of the CUSUM of square test can be computed for any given level of the test. For the sake of our study, in this chapter we have concentrated only on the power computation of the test. The power results of this test are tabulated in table 9.4 for various values of ρ and τ in table-4. The details power comparison has been discussed in section three.

The Lagrange Multiplier Test

The LM-test was first proposed by Nyblom (1989) to test for the constancy of regression coefficients, and was generalised by Hansen (1991) to test for the change in variance. In the case of a sequence of normal random variables, for testing the change in variances, the LM-test statistics can be written as,

$$LM = \frac{\sum_{k=1}^{n} \left\{ \sum_{i=1}^{k} \left[\left(X_i - \overline{X} \right)^2 - \hat{\sigma}^2 \right] \right\}^2}{n \cdot \sum_{i=1}^{n} \left[\left(X_i - \overline{X} \right)^2 - \hat{\sigma}^2 \right]^2}$$

$$\hat{\sigma}^2 = \frac{\sum_{k=1}^{n} \left(X_i - \overline{X} \right)^2}{n}$$

and where the sample mean is expressed as usual,

$$\overline{X}_i = \frac{\sum_{j=1}^{i} X_j}{i}$$

For testing H_0: $\sigma_1 = \sigma_2$ against H_1 : $\sigma_1 \neq \sigma_2$, we reject H_0 for large values of LM. The procedures of computing critical values are the same as discussed above in earlier subsections and explicitly detailed in Nybolm and Hensen's chapters.

9.3. EMPIRICAL POWER COMPARISONS

The main objective of this section is to make power comparisons between L, B, R, C and LM tests for a sample of sizes n=30. The power of the test under H1 at a given n depends on the location of the change point, τ and the size of the shift ρ where $\rho^2 = \sigma_2^2 / \sigma_1^2$. If the sequence Xl,...,Xn is inverted to the sequence Xn,...,Xl, it is easy to prove that the power at τ, ρ, is equal to that at $\tilde{n}$ τ), ρ-l , so we just make power comparisons for $\rho > 1$ We take ρ to be the solutions of the equations P($\mid Y \mid > \mid X \mid$) = 0.7,0.8,0.9, respectively. Where Y ~ N (0, ρ2), X ~ N (0, 1), the solutions are 1.4, 1.8, 2.5, respectively. Through this method, we carried out our simulations experiment. In this experiment we generated 2000 samples of size 30 and obtained the power values of the L, B, R, C and LM tests at the level =0.05 for testing H0:σl = σ2 against H1:$\sigma_1 \neq \sigma_2$ using the Monte Carlo simulation. Some numerical results of our simulation and its illustrations are given in table 9.4.

An empirical power comparison of these tests demonstrates that some of the tests have higher power when the change point τ, is in mid portion of the sequence, compared to that of values when τ is either near or the beginning or end of the sequence. When τ is in the beginning, the L and B tests is comparable to its competitors. When τ moves from the beginning to mid portion however, the L test is better, whilst when τ moves from mid portion to the end, the power of the B test is the more superior than the other two. Generally speaking, the B test proves itself to be the best. The low power values reported in table 9.4 are mainly a reflection of the lack of information regarding (the mean) μ, (the initial value of the standard deviation) $\sigma 1$, and τ (the location of the change point).

9.4. CONCLUDING REMARKS

This chapter had considered tests for testing change point problems of variances in the sequence of independent random variables of the type SMC with additional complications of unknown means. An empirical power study was conducted of five tests, to see which test held the highest power to be called an approximated uniformly most powerful test for change point testing problems. The tests studied were, the Lehmann U-statistic, a test derived using the Bayesian method, the R-test, the C-test, and the LM test. It is observed that the Bayesian test (B-test), and the CUSUM test (C-test) were marginally superior to those of the test based on the Lehmann U-statistic (L), the LM test and the R-tests. An empirical power comparison of the five tests favoured the power gain of the Bayesian method (B-test). Our findings from calculations and our comparison however suggest that the power gained among these tests is not clear, overall in relation to the parametric space under test. Therefore more work is required to understand change point problems.

APPENDIX 9A: APPROXIMATE CRITICAL VALUES OF THE L TEST

Table 9.1. Approximate critical values of the L test

N	=0.01 LL_α	L_α	α=0.05 LL_α	L_α	α=0.10 LL_α	L_α
10	2.3752	2.4945	2.0194	2.2342	1.8412	2.0788
15	2.6618	2.8125	2.2668	2.4549	2.0239	2.2615
20	2.7874	2.9619	2.3351	2.5706	2.1166	2.3576
25	2.8757	3.0840	2.4500	2.6533	2.1873	2.4380
30	2.9505	3.1202	2.4662	2.7037	2.2234	2.4774
35	3.0025	3.1769	2.4852	2.7272	2.2387	2.4860
40	3.0408	3.1933	2.5363	2.7836	2.2880	2.5289
45	3.0697	3.2128	2.5489	2.8065	2.2989	2.5562
50	3.0975	3.2948	2.5761	2.8562	2.3143	2.6085

Table 9.2 Approximate critical values of the B test

N	$B_{0.90}$	$B_{0.95}$	$B_{0.975}$	$B_{0.99}$	$B_{0.995}$
10	1.3005	1.6520	1.9452	2.2706	2.4796
15	1.2943	1.6503	1.9510	2.2887	2.5116
20	1.2913	1.6494	1.9539	2.2976	2.5297
25	1.2895	1.6489	1.9556	2.3036	2.5401
30	1.2883	1.6486	1.9567	2.3079	2.5468
35	1.2874	1.6483	1.9576	2.3110	2.5515
40	1.2868	1.6481	1.9582	2.3133	2.5550
45	1.2863	1.6480	1.9586	2.3150	2.5577
50	1.2859	1.6479	1.9590	2.3164	2.5598

Table 9.3 Approximate critical values of the R test

N	$R_{0.90}$	$R_{0.95}$	$R_{0.99}$
10	0.0387529	0.0187973	0.0039700
15	0.0311388	0.0145218	0.0026099
20	0.0267761	0.0119599	0.0020611
25	0.0248677	0.0111856	0.0019054
30	0.0214734	0.0104916	0.0018094
35	0.0213546	0.0104478	0.0016385
40	0.0198045	0.0097026	0.0015905
45	0.0190787	0.0096263	0.0013672
50	0.0178123	0.0081291	0.0012219

Table 9.4. Power values of the L, B, R, C and LM-tests

τ	ρ	L-test	B-test	R-test	C-test	LM-test
3	1.4	0.067	0.015	0.064	0.050	0.054
	1.8	0.077	0.063	0.088	0.053	0.058
	2.5	0.116	0.077	0.141	0.054	0.069
7	1.4	0.100	0.091	0.088	0.064	0.086
	1.8	0.219	0.154	0.173	0.100	0.129
	2.5	0.483	0.224	0.417	0.153	0.187
1 1	1.4	0.138	0.147	0.111	0.102	0.140
	1.8	0.332	0.272	0.259	0.234	0.243
	2.5	0.725	0.465	0.629	0.460	0.390
1 5	1.4	0.142	0.196	0.114	0.178	0.177
	1.8	0.358	0.394	0.318	0.426	0.357
	2.5	0.766	0.721	0.724	0.825	0.599
1 9	1.4	0.124	0.229	0.120	0.213	0.195
	1.8	0.323	0.458	0.338	0.473	0.392
	2.5	0.708	0.825	0.730	0.861	0.674
2 3	1.4	0.095	0.201	0.111	0.193	0.168
	1.8	0.217	0.407	0.279	0.413	0.324
	2.5	0.551	0.760	0.654	0.774	0.591
2 7	1.4	0.054	0.123	0.073	0.110	0.103
	1.8	0.094	0.252	0.162	0.228	0.161
	2.5	0.191	0.487	0.404	0.493	0.286

APPENDIX 9B: PROOF OF LEMMA 9.1

Without loss of generality, let $X_i \sim N(0,1), i = 1,...,n$. We know

$$D(U_{k,m-k}) = \left\{ \binom{k}{2}\binom{n-k}{2} \right\}^{-1} \cdot \sum_{c_1=0}^{2} \sum_{c_2=0}^{2} \binom{2}{c_1}\binom{2}{c_2}\binom{k-2}{2-c_1}\binom{n-k-2}{2-c_2}\sigma_{c_1 c_2}^2$$

where

$$\sigma_{c_1 c_2} = D\phi(X_1,...,X_{c_1}; X_{k+1}, X_{k+c_2})$$
$$= D\left\{ E\left[h\left(X_1, X_2; X_{k+1}, X_{k+2} \right) \Big| X_1,...,X_{c_1}; X_{k+1}, X_{k+c_2} \right] \right\}$$

Obviously, $\sigma_{00}^2 = 0$ and $\sigma_{22}^2 = \dfrac{1}{4}$.

$$P\left(|X_{k+1} - X_{k+2}| > |x - X_2| \right) = P\left(\frac{Y_1}{Y} < 2 \right)$$

where $Y_1 \sim \chi^2(1, x^2/2) Y^2 \sim \chi^2(1)$. Hence,

$$\phi(x) = \sum_{r=0}^{\infty} \exp\left(-\frac{x^2}{2} \right) \cdot \frac{x^{2r}}{2^r \cdot r!} \cdot B(r)$$

where

$$B(r) = \frac{\Gamma(1+r)}{\Gamma\left(\dfrac{1}{2}+r\right) \cdot \Gamma\left(\dfrac{1}{2}\right)} \int_0^{\frac{2}{3}} u^{r+\frac{1}{2}-1} \cdot (1-u)^{-\frac{1}{2}} du$$

Hence

$$\sigma_{10}^2 = \sigma_{01}^2 = D\phi(X_1) = E\left[\phi(X_1)\right]^2 - 0.25$$

$$= \sum_{r=0,s=0}^{\infty}\sum_{s=0}^{\infty} \frac{B(r)B(s)}{6^{r+s}\, r!s!\sqrt{3}}\left[2(r+s)-1\right]!! - 0.25$$

$$= 0.0138$$

$$\phi(x,y) = P\left(\left|X_{k+1} - X_{k+2}\right| < \left|x-y\right|\right) = 2\cdot\int_0^{\frac{|x-y|}{2}} \frac{1}{\sqrt{2\pi}}\exp\left(\frac{t^2}{2}\right)dt$$

$$= \frac{2}{\sqrt{\pi}}\exp\left(-\frac{(x-y)^2}{4}\right)\cdot\sum_{r=1}^{\infty}\frac{2^{r-1}}{(2r-1)!!}\left(\frac{|x-y|}{2}\right)^{2r-1}$$

Hence

$$\sigma_{20}^2 = \sigma_{02}^2 = D\phi(X_1,X_2) = E\left[\phi(X_1,X)\right]^2 - 0.25$$

$$= \sum_{r=1}^{\infty}\sum_{s=1^*}^{\infty} \frac{\left[2(r+s)-3\right]!!}{\pi(2r-1)!!(2s-1)!!3^{r+s-\frac{1}{2}}} - 0.25$$

$$= 0.0833$$

$$\phi(x,y) = P\left(\left|x - X_{k+2}\right| < \left|y - X_2\right|\right)$$

$$= \sum_{r=0s}^{\infty}\sum_{=0}^{\infty} \frac{\exp\left(-\dfrac{x^2+y^2}{2}\right)}{r!s!}\left(\frac{x^2}{2}\right)r\cdot\left(\frac{y^2}{2}\right)s\cdot B(r,s)$$

where

$$B(r,s) = \frac{\Gamma(r+s-1)}{\Gamma\left(r+\dfrac{1}{2}\right)\Gamma\left(s+\dfrac{1}{2}\right)}\cdot\int_0^{\frac{1}{2}} u^{s-\frac{1}{2}}.(1-u)^{r-\frac{1}{2}}du$$

Hence

$$\sigma_{11}^2 = D\phi(X_1, X_{k+1}) = E\left[\phi(X_1, X_{k+1})\right]^2 - 0.25$$

$$= \sum_{r_1=0}^{\infty} \sum_{s_1=0}^{\infty} \sum_{r_2=0}^{\infty} \sum_{s_2=0}^{\infty} \frac{B(r_1, s_1) \cdot B(r_2, s_2) \cdot \left[2(r_1 + r_2) - 1\right]!! \left[2(s_1 + s_2) - 1\right]!!}{3 \cdot r_1! \, s_1! \, s_2! \, 6^{r_1 + s_1 + r_2 + s_2}} - 0.25$$

$$= 0.0209$$

$$\phi(x, y, z) = p\left(\left|x = X_{k+2}\right| > \left|y - z\right|\right)$$

$$\sum_{r=0}^{\infty} \frac{\exp\left(-\dfrac{x^2}{2}\right)}{r!} \cdot \left(\frac{x^2}{2}\right)^{r(y-z)^2} \int_0^{\infty} \frac{\left(\dfrac{1}{2}\right)^{r+\frac{1}{2}}}{\Gamma\left(r + \dfrac{1}{2}\right)} u^{r - \frac{1}{2}} \exp\left(-\frac{u}{2}\right) du$$

$$\sum_{r=0}^{\infty} \frac{\exp\left(-\dfrac{x^2}{2}\right)}{r!} \cdot \left(\frac{x^2}{2}\right)^2 \frac{2^s \cdot (2r-1)!!}{\Gamma\left(r + \dfrac{1}{2}\right) \cdot \left[2(r+s) - 1\right]!!}$$

$$\exp\left(\frac{(y-z)^2}{2}\right)\left(\frac{\left|y-z\right|}{\sqrt{2}}\right)^{2(r+s)-1}$$

Hence

$$\sigma_{12}^2 = \sigma_{21}^2 = D\phi(X_1, X_2, X_{k+1}) = E\left[\phi(X_1, X_2, X_{k+1})\right]^2 - 0.25$$

$$= \sum_{r_1=0}^{\infty} \sum_{s_1=1}^{\infty} \sum_{r_2=0}^{\infty} \sum_{s_2=1}^{\infty}$$

$$= \frac{\left[2(r_1 + r_2) - 1\right]!!(2.r_1 - 1)!!(2.r_2 - 1)!!\left[2(r_1 + s_1 + r_2 + s_2) - 3\right]!!}{r_1! r_2! \left[2(r_1 + s_1) - 1\right]!! \left[2(r_2 + s_2) - 1\right]!! \Gamma\left(r_2 + \dfrac{1}{2}\right)}$$

$$\frac{2^{s_1 + s_1 + r_1 - r_2}}{3^{r_1 + r_2 + \frac{1}{2}} 5^{r_1 + r_2 + s_1 + s_2 - \frac{1}{2}}} - 0.25$$

$$= 0.1004$$

This proofs the lemma (9.1).

10. CONCLUDING REMARKS

In this book several aspects of hypothesis testing procedures associated with the linear regression model have been considered and then applied to the multistage sample clustered data. The principal aim of this book was to develop enhanced optimal tests (i.e. LB, LMMP, BO and PO) to ascertain their size and power properties, and to empirically compare the powers of the existing tests with those of the new optimal tests and asymptotic LM tests. At the beginning of this study, optimal testing procedures were seen to be very tedious and cumbersome in choosing a power curve that could dominate the power of other testing procedures. However, the use of computer technology and the advances in numerical methods assisted with the choice of a test procedure that was most powerful for testing cluster effects change point problems and hence provides a more accurate decision in hypothesis testing. The key to success was in understanding how to apply these procedures to the issues raised in chapter one.

In chapter two, literature was reviewed to increase the understanding of the reader on the issues and problems connected with sample surveys, regression analysis, optimal testing and existing tests for testing cluster effects. Chapter two also laid down the conditions under which the study in this book would be undertaken, with a unified regression model developed in the process, which could be used by both econometricians and statisticians in practice. In the remaining of chapter two, the power properties of the optimal tests for testing cluster effects were discussed, with various models was explored.

The purpose of chapter three whereas was to develop BO, PO and LMMP tests for testing intra cluster correlation coefficients in the two- and three-stage

SSMC distribution models. Using the numerical iterative procedure suggested by King (1987) and Davies and Harte (1987), the PE was computed and compared to the powers of the SenGupta's LB test, and the two versions of the BO tests, namely, $r_{0.5}$ and $r_{0.8}$ at the five percent level of significance. It was found that a BO test for the two-stage SSMC model was superior to SenGupta's LB test, and also that it is approximately UMP. In chapter four, the analogous tests for the SMC distribution model were considered, with it found that the power and the critical values of the POI and LMMPI test for the case of the two-stage SMC model were easily obtainable from the standard F distribution, and that these tests were also UMPI. The POI and LMMPI tests for the case of the three-stage SMC model were derived and applied, with the POI test for the more complicated problem of testing subcluster correlation coefficient in the presence of main cluster correlation.

The main purpose of chapter five was to consider a 2SLR model whose disturbances follow SMC distributions, and to construct a POI test for testing cluster effects. The PE was computed and then compared with the powers of the POI and existing tests from the King and Evans studies. It was found through this that the powers of the POI test were marginally better than those of the existing tests, and that for some selected cluster sizes, the POI test was approximately UMPI, and its approximate critical values, c_{α}^{*} are obtainable from the standard F distribution. It was observed that in 87 per cent of the cases, the estimated sizes (based on c_{α}^{*}) were more than half of the nominal size. The LMMPI test was developed for testing multiparameters, and was found to be equivalent to Deaton and Irish's LM1 test.

In chapter six, the 3SLR model was considered and its content was mainly concentrated on the following two issues. Firstly, the LMMPI and POI tests, which were constructed for testing whether the cluster/subcluster effects were equal to zero, or were positive. An evaluation of the power showed that the LMMPI test has quite good power for the data sets that were used in this experiment, which supports the usefulness of the study and the test. Secondly, a test for testing intra-subcluster equicorrelation in the presence of intra-cluster equicorrelation was constructed, with an exact POI test and asymptotic LM tests were developed. The exact sizes and powers of the POI tests were computed using numerical iterative procedures, whilst the sizes and power of the LM tests were estimated using Monte Carlo simulation techniques. The estimated sizes of the LM1 test for the 100 per cent of the cases considered in this experiment differed significantly than the nominal size of five percent. The sizes of the LM1 test have the tendency to increase with an increase in the sample size, whereas the sizes of the LM2 test instead tended decreases. Generally, the power pattern of the POI and LM1 tests were similar, as it was the case of 2SLR model in chapter five, i.e. the power of the PO1 test is marginally better than that of the LM1 test.

The optimal testing procedures developed in chapter seven were concerned with testing for a serial correlation in a large number of small samples, by generalizing the work of Cox and Solomon (1986, 1988). Various tests related to LNSG models were derives and then linked with prevision chapters. Chapter eight whereas developed and illustrated efficient estimation procedures for the stochastic coefficients model, cluster-wise heteroscedasticity and intra cluster correlation, based on two-stage-clustered survey data. In chapter nine, we conducted five tests, namely L, B, R, C and LM tests, for testing change point problems associated with SMC and SSMC type models. Critical values and power computation was simulated. Results of the power comparison support the power gain of the B-test. However a more in-depth study needs to be in this area.

In conclusion, it can be said that for testing cluster effects, if a problem satisfies the conditions for the use of optimal testing, one should be encouraged to use such a procedure. In a situation where no UMP test exists, the correct choice from among optimal tests is important. In choosing a test, one should always search for a test which through its power dominates others in the alternative parameter space. This book further strengthens the claim of the superiority of the PO test in the area of testing for cluster effects, with the powers of the PO tests shown to be superior than those of the other existing tests. The main constraint on the generalised use of these tests was that they required a prior knowledge of the signs of the parameters under testing. In reality one cannot and could not conclude, despite this exercise, that optimal tests are the perfect tests in all circumstances. There is still a need for a great deal of research to be undertaken in the area, but as shown in this book, optimal tests are worthy of attention.

ACKNOWLEDGMENTS

Whatever merits of this book may be there are in large measures due to my parents, teachers and colleagues I owe a great deal for their help, support, and encouragement during the preparation of the first version of this book in 1995. Three additional chapters are from my published work coauthored with Wu Ping, Linxing Zhu and Jinglong Wang. I take this opportunity to thank them all.

I also appreciate the patience, understanding and support of my wife and my children, Maria, Osama and especially the youngest and the favourite Taha, who are invaluable parts of my life, and whose company I have been blessed with.

I would like to express my appreciation to Tulsi Rajyaguru, my Research Assistant and Suren Basov for their support, companionship and constructive comments. Partial funding from ARC Discovery grant is acknowledged. I would also like to thank Maya Columbus, Editor in Nova Science Publishers for taking up with this project.

BIBLIOGRAPHY

Ali, M. (2002), Historical ex-post and ex-ante forecasts for three kinds of pulse in Bagladesh: Masu, Gram and Khesuri, *Unpublished PhD dissertation* Jahangir Nagar University, Bangladesh.

Amemiya, T. (1971), 'The Estimation of Variances in a Variance-Components Model', *International Economic Review*, 12, pp. 1-13.

Anderson, T.W. (1984), *An Introduction to Multivariate Statistical Analysis*, John Wiley, New York.

Andrews, D.W.K. (1993). Tests for parameter instability and structural change with unknown change point, *Econometrica*, 61, 821-856.

Ansley, C.F. (1979), 'An Algorithm for the Exact Likelihood of a Mixed Autoregressive-Moving Average Process', *Biometrika*, 66, pp. 59-65.

Antoch, J., and Huskova, M (1992). Some M-tests for detection of a change in the linear models, Proceedings of the 4th Prague Symposium on Asymptotic Statistics, Charles University Press, Praha, pp. 123-136.

Ara, I. and King, M.L. (1993), 'Marginal Likelihood Based Tests of Regression Disturbances'. Paper presented at the 1993 meeting of the Australasian meeting of the Econometrics Society, University of Sydney, Sydney, Australia.

Balestra, P. and Nerlove, M. (1966), 'Pooling Cross Section and Time Series Data in the Estimation of a Dynamic Model The Demand for Natural Gas', *Econometrica*, 34, pp. 585-612.

Belisle, P. Joseph, L., MacGibbon, B., Wolfson, D. B. and du Berger, R. (1998). "Change point analysis of neuron spike train data" *Biometrika,* 54, 107-117.

Baltagi, B.H. (1981), 'Pooling: An Experimental Study of Alternative Testing and Estimation Procedures in a Two-Way Error Component Model', *Journal of Econometrics*, 17, pp. 21-49.

Baltagi, B.H., Chang, Y.J. and Li, Q. (1992), 'Monte Carlo Results on Several New and Existing Tests for the Error Component Model', *Journal of Econometrics*, 54, pp. 95-120.

Baltagi, B.H. and Li, Q. (1991), 'A Joint Test for Serial Correlation and Random Individual Effects, ' *Statistics and Probability Letters*, vol 11, pp. 277-280.

Baltagi, B.H. and Raj, B. (1992), 'A Survey of Recent Theoretical Developments in the Econometrics of Panel Data, ' *Empirical Economics*, vol 17, pp. 85-109.

Bhatti, M.I. (1992), 'Optimal Testing for Serial Correlation in Large Number of Small Samples', *Biometrical Journal*, Vol. 34, pp. 57-67.

Bhatti, M.I. (1993), 'Efficient Estimation of Random Coefficient Models Based on Survey Data'. *Journal of Quantitative Economics*, Vol. 9, No. 1, pp. 99-110.

Bhatti, M.I. (1994), 'Optimal Testing for Equicorrelated Linear Regression Models', to appear in Statistical Papers.

Bhatti, M.I. (1995a), 'A Ump-Invariant Test: An Example', *Journal of Applied Statistical Science*, Vol. 2, No.4.

Bhatti, M. I. (1995b). Testing regression models based on sample survey data, Avebury, Ashgate publiing, UK.

Bhatti, M. I (2000), On Optimal Testing for Equicorrelated LinearRegression Models, *Statistical Papers, Vol. 36(4), pp. 299-312.*

Bhatti, M. I. (2001) Econometric Analysis of Marketing Potential ofOICCountries: Some Facts under Global Economy, with M. Z.Hossain, M. Z. Hoq, *Managerial Auditing, Vol. 20 (2), 2005.*

Bhatti, M. I. (2004), On Cluster Effects in Mining Complex EconometricData. In Statistical Data Mining and Knowledge Discovery, *H.   Bozdogan Ed.),Chapter 23, pp.387-399, Chapman and Hall/CRC, USA.*

Bhatti, M.I. and King, M.L. (1990), 'A Beta-Optimal Test of The Equicorrelation Coefficient', *Australian Journal of Statistics*, 32, pp. 87-97.

Bhatti, M. I., Al-Shanfari, H., and Hossain, M. Z. (2006). *Econometrics Analysis of Model Testing and Model Selection*, Ashgate Publishing, UK

Bhatti, M. I. Wang, J. (1998). Power comparison of some tests for testing change point, *Statistika,* Anno, LVIII, n. 1, pp. 127-137.

Bhatti, M. I and Wang, J. (2005). Tests and confidence intervals for Change point in scale of two-parametric exponential distributions, with J Wang, Journal of Applied Statistical Sciences, Vol. 14(1), pp. 45-57.

Bowley, A.L. (1906), 'Address to the Economic Science and Statistics Section of the British Association for the Advancement of Science', Journal of the Royal Statistical Society, 69, pp. 548-557.

Breusch, T.S. (1980), 'Useful Invariance Results for Generalized Regression Models', Journal of Econometrics, 13, pp. 327-340.

Breusch, T.S., and Pagan, A.R. (1980), 'The Lagrange Multiplier Test and its Applications to Model Specification in Econometrics', Review of Economic Studies, 47, pp. 239-253.

Brooks, R. (1993), 'Alternate Point Optimal Tests for Regression Coefficient Stability', Journal of Econometrics, 57, pp. 356-376.

Brook and King (1994), Hypothesis testing of varying coefficient regression models: Procedures and applications, Pakistan Journal of Statistics, 10, 301-357.

Brown, R. L., Durbin, J. and Evans, J. M. (1975). Techniques for testing the constancy of regression relationships over time, Journal of the Royal Statistical Society, B, 37, 149-163.

Buse, A. (1980), 'Tests for Additive Heteroscedasticity: Some Preliminary Results', paper presented to the Fourth World Congress of the Econometric Society, Aix-en-Provence.

Campbell, C. (1977), 'Properties of Ordinary and Weighted Least Squares Estimates of Regression Coefficients for Two-Stage Samples', in Proceedings of the Social Statistics Section, American Statistical Association, pp. 800-805.

Chernoff, H. and Zacks, S. (1964), Estimating the current mean of a normal distribution which is subject to changes over-time, "Annals of Mathematical Statistics", 35, pp. 999-1018

Christensen, R. (1984), 'A Note on Ordinary Least Squares Methods for Two-Stage Sampling', Journal of the American Statistical Association, 79, pp. 720-721.

Christensen, R. (1986), 'Methods for the Analysis of Cluster Sampling Models', Technical Report, No. 7, pp. 22-86, Montana State University, Department of Mathematical Sciences.

Christensen, R. (1987a), 'The Analysis of Two-stage Sampling Data by Ordinary Least Squares', Journal of the American Statistical Association, 82, pp. 492-498.

Christensen, R. (1987b), *Plane Answers to Complex Questions: The Theory of Linear Models*, Springer-Verlag, New York.

Cochran, W.G. (1942), 'Sampling Theory When the Sampling Units are of Unequal Sizes', *Journal of the American Statistical Association*, 37, pp. 199-212.

Cochran, W.G. (1946), 'Relative Accuracy of Systematic and Stratified Random Samples for a Certain Class of Population', *Annals of Mathematical Statistics*, 17, pp. 164-177.

Cochran, W.G. (1953), *Sampling Techniques,* 1st edition, 2nd edition (1963), John Wiley, New York.

Cox, D.R. (1961), 'Tests of Separate Families of Hypotheses', Proceedings of the Fourth Berkeley Symposium on Mathematical Statistics and Probability, 1, University of California Press, Berkeley, pp. 105-123.

Cox, D.R. (1962), 'Further Results on Tests of Separate Families of Hypotheses', *Journal of the Royal Statistical Society* B, 24, pp. 405-424.

Cox, D.R. and Hinkley, D.V. (1974), *Theoretical Statistics*, Chapman and Hall, London.

Cox, D.R. and Solomon, P.J. (1986), 'Analysis of Variability with Large Numbers of Small Samples', *Biometrika*, 73, pp. 543-554.

Cox, D.R. and Solomon, P.J. (1988), 'On Testing for Serial Correlation in Large Numbers of Small Samples', *Biometrika*, 75, pp. 145-148.

Crowder, M.J. (1985), 'A Distributional Model for Repeated Failure Time Measurements', *Journal of the Royal Statistical Society*, B, 47, pp. 447-452.

Csorgo, M., and Horvath, L. (1982). Nonparametric methods for the change point problem, Handbook of Statistics, vol. 7 (P.R.Krishnaiah and C.R.Rao, eds.) J. Wiley, New York, pp. 403-425.

Davern, M., Jones Jr, A., Lepkowski, J., Davidson, G., and Blewett, L. A. (2007), 'Estimating Regression Standard Errors with Data from the Current Population Survey's Public Use File', *Inquiry - Excellus Health Plan*, 44, pp 211- 224.

Davies, R.B. (1969), 'Beta-Optimal Tests and an Application to the Summary Evaluation of Experiments', *Journal of the Royal Statistical Society*, B, 31, pp. 524-538.

Davies, R.B. (1980), 'Algorithm AS155. The Distribution of a Linear Combination of $\chi 2$ Random Variables', *Applied Statistics*, 29, pp. 323-333.

Davies, R.B. and Harte, D.S. (1987), 'Tests for Hurst Effect', *Biometrika*, 74, pp. 95-101.

Deaton, A. and Irish, M. (1983), 'Block Effects in Regression Analysis Using Survey Data', unpublished manuscript.

Deming, W.E. (1950), *Some Theory of Sampling*, Wiley, New York.

Desai, M., and Begg, M. (2008), 'A Comparison of Regression Approaches for Analyzing Clustered Data', *American Journal of Public Health*, Vol. 98, No. 8, pp 1425- 1429.

Dickens, W.T. (1990), 'Error Components in Grouped Data: Is It Ever Worth Weighting?' *The Review of Economics and Statistics*, vol. 72, pp. 328-333.

Dorfman, A.H. (1993), 'A Comparison of Design-Based and Model-Based Estimators of the Finite Population Distribution Function, *Australian Journal of Statistics*, Vol. 35(1), pp. 29-41.

Dunlap, W. P., Xin, X., and Myers, L. (2004), 'Computing aspects of power for multiple regression', *Behavior Research Methods*, Vol. 36, No. 4, pp 695- 701.

Efron, B. (1975), 'Defining the Curvature of a Statistical Problem (with Applications to Second Order Efficiency)', *Annals of Statistics*, 3, pp. 1189-1242.

Efron, B. (1979), "Bootstrap methods: another look at the jackknife", *Annals of Statistics, 7:* 1-26.

Efron, B. (1982), "The Jackknife, the Bootstrap, and other resampling plans", (Philadelphia: Society for Industrial and Applied Mathematics).

Evans, M.A. and King, M.L. (1985), 'A Point Optimal Test for Heteroscedastic Disturbances', *Journal of Econometrics*, 27, pp. 163-178.

Farebrother, R.W. (1980), 'Algorithm AS153: Pan's Procedure for the Tail Probabilities of the Durbin-Watson Statistic', *Applied Statistics*, 29, pp. 224-227 and 30, pp. 189.

Ferguson, T.S. (1967), *Mathematical Statistics A Decision Theoretic Approach*, Academic Press, New York.

Fisher, R.A. (1925), 'Theory of Statistical Estimation', Proceedings of the Cambridge Philosophical Society, 122, pp. 700-725.

Fuller, W.A. (1975), 'Regression Analysis for Sample Surveys', Shankhya, C, 37, pp. 117-132.

Fuller, W.A. and Battese, G.E. (1973), 'Transformations for Estimation of Linear Models with Nested Error Structures', *Journal of the American Statistical Association*, 68, pp. 626-632.

Fuller, W.A. and Battese, G.E. (1974), 'Estimation of Linear Models with Cross-Error Structure', *Journal of Econometrics*, 2, pp. 67-78.

Geisser, S. (1963), 'Multivariate Analysis of Variance for a Special Covariance Case', *Journal of the American Statistical Association*, 58, pp. 660-669.

Godambe, V.P. (1966), 'A New Approach to Sampling from Finite Populations', *Journal of the Royal Statistical Society*, B, 28, pp. 310-328.

Godfrey, L.G. (1988), *Misspecification Tests in Econometrics: The Lagrange Multiplier Principle and other Approaches*, Cambridge University Press, Cambridge.

Goldfeld, S.M. and Quandt, R.E. (1965), 'Some Tests of Heteroscedasticity', *Journal of the American Statistical Association*, 60, pp. 539-547.

Greenwald, B.C. (1983), 'A General Analysis of the Bias in the Estimated Standard Errors of Least Squares Coefficients', *Journal of Econometrics*, 22, pp. 323-338.

Graybill, F.A. (1969), *Introduction to Matrices with Applications in Statistics*, Wadsworthy, Belmont, California.

Halperin, M. (1951), 'Normal Regression Theory in the Presence of Intra-Class Correlation', *The Annals of Mathematical Statistics*, 22, pp. 573-580.

Hansen, B.E. (1991). Lagrange multiplier test for parametric instability in nonlinear models, University of Rochester Working Chapter

Hansen, M.H. and Hurwitz, W.N. (1949), 'On the Determination of the Optimum Probabilities in Sampling', *Annals of Mathematical Statistics*, 20, pp. 426-432.

Hansen, M.H., Hurwitz, W.N. and Madow, W.G. (1953), *Sampling Survey Methods and Theory*, Volumes I and II, John Wiley, New York.

Hartley, H.O. and Rao, J.N.K. (1968), 'A New Estimation Theory for Sample Survey', *Biometrika*, 55, pp. 547-557.

Hartley, H.O. and Rao, J.N.K. (1969), 'A New Estimation Theory for Sample Survey II, ' *In New Developments in Survey Sampling*, N.L. Johnson and H. Smith (Eds.), pp. 147-169, Wiley Interscience, New York,.

Harvey, A.C. and Phillips, G.D.A. (1974), 'A Comparison of the Power of Some Tests for Heteroscedasticity in the General Linear Model', *Journal of Econometrics*, 2, pp. 307-316.

Harrison, M.J. (1980), 'The Small Sample Performance of the Szroeter Bounds Tests for Heteroscedasticity and a Simple Test for Use When Szroeter's Test is Inconclusive', *Oxford Bulletin of Economic and Statistics*, 42, pp. 235-250.

Harrison, M.J. and McCabe, B.P.M. (1979), 'A New Test for Heteroscedasticity Based on Ordinary Least Squares Residuals', *Journal of the American Statistical Association,* 74, pp. 494-499.

Hausman, J.A. (1978), 'Specification Tests in Econometrics', *Econometrica*, 46, pp. 1251-1271.

Hawkins, D. M. (1977). "Testing a sequence of observations for a shift in location", *J. Am. Statist. Assoc.*, 72,180-186.

Hillier, G.H. (1987), 'Classes of Similar Regions and Their Power Propertiesfor Some Econometric Testing Problems, ' *Econometric Theory*, pp. 1-44.

Hinkley, D. V. (1969). Inference about the intersection in two-phaseregression, *Biometrika,*56, 495-504.

Hinkley, D. V. (1970). "Inference about the change point in a sequence ofrandom variables", *Biometrika*, 57,1-17.

Hinkley, D. V. (1971). Inference about a change point from cumulative sumtests, *Biometrika*, 58, 509-523.

Hinkley, D. V. and Hinkley, E. A. (1970). "Inference about the changepoint in a sequence of binomial random variable", *Biometrika*, 57, 477-488.

Holt, D. and Scott, A.J. (1981), 'Regression Analysis Using Survey Data', *The Statistician*, 30, pp. 169-178.

Honda, Y. (1985), 'Testing the Error Components with Non-normal Disturbances', *Review of Economic Studies*, 52, pp. 681-690.

Honda, Y. (1989), 'On the Optimality of Some Tests of the Error Covariance Matrix in the Linear Regression Model', *Journal of the Royal Statistical Society*, Series B, Vol. 51, pp. 71-79.

Honda, Y. (1991) 'A Standardized Test for the Error Components Model with the Two-way Layout', *Economic Letters*, 37, pp.125-128.

Hoque, A. (1988), 'Farm Size and Economic-Allocative Efficiency in Bangladesh Agriculture', *Applied Economics*, 20, pp. 1353-1368.

Hoque, A. (1991), 'An Application and Test for a Random Coefficient Model in Bangladesh Agriculture', *Journal of Applied Econometrics*, 6, pp. 77-90.

Horton, N. J., and Lipsitz, S. R. (2001). 'Multiple Imputation in Practice: Comparison of Software Packages for Regression Models with Missing Variables', *The American Statistician*, Vol. 55, No. 3, pp 244 – 254.

Hossain, M. (1990), "Natural Calamities, Instability in Production and Food Policy in Bangladesh", The Bangladesh Development Studies, Vol. XVIII, No. 4, 33-54

Hsiao, C. (1986), *Analysis of Panel Data*, Cambridge University Press.

Hsu, D. A. (1979). "Detecting, shifts of parameter in gamma sequences with applications to stock price and air traffic fow analysis", *J. Am. Statist. Assoc.*, 74,31-40.

Huskova, M. and Sen, P. K. (1989). Nonparametric tests for shift and change in regression at an unknown time point, *Statistical Analysis and*

Forecasting of Economic Structural Change (Hackl P., ed.), Springer-Verlag, New York, pp.71-85.

IMSL Math/Library (1989), *User's Manual*, Softcover Edition 1.1, Houston ISML Inc., U.S.A.

Isaacson, S.L. (1951), 'On the Theory of Unbiased Tests of Simple Statistical Hypotheses Specifying the Values of Two or More Parameters', *Annals of Mathematical Statistics*, 22, pp. 217-234.

Johnson, N. L. and Kotz, S. (1972). Distributions in Statistics: Continuous Multivariate Distributions, John Wiley and Sons, New York.

Kander, Z. and Zacks, S. (1966). "Test procedures for possible changes inparameters of statistical distributions occurring at unknown timepoints", *Ann. Math. Statist.*, 37,1196-1210.

Kiaer, A.N. (1895), 'Observation et Experiences Concernant Des Denombrements Representatifs', Discussion Appears in Liv. 1, XC111-XCV11. Bulletin International Statistical Institute, 9, Liv. 2, pp. 176-183.

King, M.L. (1981), 'The Alternative Durbin-Watson Test: An Assessment Durbin and Watson's Choice of Test Statistic', *Journal of Econometrics*, 17, pp. 51-66.

King, M.L. (1982), 'A Locally Optimal Bounds Test for Autoregressive Disturbances', a paper presented at the European meeting of the Econometric Society, Dublin.

King, M.L. (1983), 'Testing for Moving Average Regression Disturbances', *Australian Journal of Statistics*, 25, pp. 23-34.

King, M.L. (1985a), 'A Point Optimal Test for Autoregressive Disturbances', *Journal of Econometrics*, 27, pp. 21-37.

King, M.L. (1985b), 'A Point Optimal Test for Moving Average Regression Disturbances', *Econometrics Theory*, 1, pp. 211-222.

King, M.L. (1986), 'Efficient Estimation and Testing of Regressions with a Serially Correlated Error Component', *Journal of Quantitative Economics*, 2, pp. 231-247.

King, M.L. (1987a), 'Testing for Autocorrelation in Linear Regression Models: A Survey'. In M.L. King and D.E.A. Giles, eds., *Specification Analysis in the Linear Regression Model*. Rutledge and Kegan Paul, London, pp. 19-73.

King, M.L. (1987b), 'Towards a Theory of Point Optimal Testing', *Econometric Reviews*, 6, pp. 169-218.

King, M.L. (1989), 'Testing for Fourth-Order Autocorrelation in Regression Disturbances when First-Order Autocorrelation is Present', *Journal of Econometrics*, 41, pp. 284-301.

King, M.L. and Evans, M.A. (1986), 'Testing for Block Effects in Regression Models Based on Survey Data', *Journal of the American Statistical Association*, 81, pp. 677-679.

King, M.L. and Evans, M.A. (1988), 'Locally Optimal Properties of the Durbin-Watson Test, ' *Econometric Theory*, 4, pp. 509-516.

King, M.L. and Hillier, G.H. (1985), 'Locally Best Invariant Tests of the Error Covariance Matrix of the Linear Regression Model', *Journal of the Royal Statistical Society*, B, 47, pp. 98-102, *Econometrics*, 25, pp. 35-48.

King, M.L. and Giles, D.E.A. (1984), 'Autocorrelation and Pre-Testing in the Linear Model Estimation, Testing and Prediction', *Journal of Econometrics*, 25, pp. 35-48.

King, M.L. and Skeels, C.L. (1984), 'Joint Testing for Serial Correlation and Heteroscedasticity in the Linear Regression Model', Paper presented at the Australasian Meeting of the Econometrics Society, Sydney.

King, M.L. and Wu, X.P. (1990), 'Locally Optimal One-Sided Tests for Multiparameter Hypothesis'. Paper presented at the Sixth World Congress of the Econometric Society, Barcelona, Spain.

Kish, L. (1965), *Survey Sampling*, John Wiley, New York.

Kish, L. and Frankel, M.R. (1974), 'Inference from Complex Samples', *Journal of the Royal Statistical Society*, B, 36, pp. 1-37.

Kloek, T, (1981), 'OLS Estimation in a Model Where a Microvariable is Explained by Aggregates and Contemporaneous Disturbances are Equicorrelated', *Econometrica*, 49, pp. 205-207.

Koerts, J. and Abrahamse, A.P.J. (1969), *On the Theory and Application of the General Linear Model*. Rotterdam University Press, Rotterdam .

Konijn, H.S. (1962), 'Regression Analysis in Sample Surveys', *Journal of the American Statistical Association*, 57, pp. 590-606.

Körösi, G., Mátyás, L. and Székely, I. (1992), *Practical Econometrics*. Avebury, Aldershot, U.K.

Krämer, W. and Sonnberger, H. (1986), *The Linear Regression Model Under Test*, Physica-Verlag, Heidelberg.

Krishnaiah, P. R., and Miao, B. Q. (1988). Review about estimation of change points, Handbook of Statistics, vol. 7. (P.R.Krishnaiah and C.R.Rao, eds) Elsevier, Amsterdam, pp. 375-402.

Lehmann, E. L. (1951). "Consistency and unbiasedness of certain non-parametric tests", *Ann. Math. Statist,* 22,165-179

Lehmann, E.L. (1959), *Testing Statistical Hypothesis*, 1st edition (2nd edition, 1986), John Wiley, New York.

Lehmann, E. L. (1986). Testing Statistical Hypotheses (2nd Edition), John Wiley and Sons, New York.

Lehmann, E.L. and Stein, C. (1948), 'Most Powerful Tests of Composite Hypothesis, I. Normal Distributions', *Annals of Mathematical Statistics*, 19, pp. 495-515.

Levin, B. and Kline, J. (1985). "The CUSUM test of homogeneity with application in spontaneous abortion epidemiology" *Statistics in Medicine*, 4, 469-488.

Luo, X. Turnbull, B. W. and Clark, L. C. (1997). "Likelihood ratio tests for a change point with survival data" *Biometrika*, 84, 555-565.

Maddala, G.S. (1971), 'The Use of Variance Components Models in Pooling Cross Section and Time Series Data', *Econometrica*, 39, pp. 341-358.

Maddala, G.S. (1977), *Econometrics*. Kogakusta, McGraw-Hill.

Madow, W.G. and Madow, L.H. (1944), 'On the Theory of Systematic Sampling', *Annals of Mathematical Statistics*, 20, pp. 333-354.

Magnus, J.R. (1978), 'Maximum Likelihood Estimation of the GLS Model with Unknown Parameters in the Disturbance Covariance Matrix', *Journal of Econometrics*, 7, pp. 281-312.

Maguire, B. A., Pearson, E. S. and Wynn, A. H. A . (1952). "The time intervals between industrial accidents", *Biometrika*, 38, 168-80.

Martin, R.S., Reinsch, C. and Wilkinson, J.H. (1968), 'Householder's Tridiagonalization of a Symmetric Matrix', *Numerische Mathematik* 11, pp. 181-195.

Mátyás, L. and Sevestre, P. (1992), *The Econometrics of Panel Data*, Kluwer Academic Publishers.

Moulton, B.R. (1986), 'Random Group Effects and the Precision of Regression Estimates', *Journal of Econometrics*, 32, pp. 385-397.

Moulton, B.R. (1990), 'An Illustration of a Pitfall in Estimating the Effects of Aggregate Variables on Micro Units', *The Review of Economics and Statistics*, 72, pp. 334-338.

Moulton, B.R. and Randolph, W.C. (1989), 'Alternative Tests of the Error Components Model', *Econometrica*, 57, pp. 685-693.

Newey, W.K. (1985), 'Maximum Likelihood Specification Testing and Conditional Moment Tests', *Econometrica*, 53, pp. 1047-1070.

Neyman, J. (1934), 'On the Two Different Aspects of the Representative Method: The Method of Stratified Sampling and the Method of Purposive Selection', *Journal of the Royal Statistical Society*, 97, pp. 559-606.

Neyman, J. (1935), 'Sur la Vérification des Hypothéses Statistiques Composées', Bulletin de la Société Mathematique de France, 36, pp. 346-366.

Neyman, J. and Pearson, E.S. (1936), 'Contributions to the Theory of Testing Statistical Hypothesis. I. Unbiased Critical Regions of Type A and Type A_1', Statistical Research Memoirs, 1, pp. 1-37.

Neyman, J. and Pearson, E.S. (1938), 'Contributions to the Theory of Testing Statistical Hypotheses II', Statistical Research Memoirs, 2, pp. 25-37.

Neyman, J. and Scott, E.L. (1967), 'On the Use of C($\Box$),Optimal Tests of Composite Hypotheses', Bulletin of the International Statistical Institute, 41, pp. 477-497.

Nimon, K., Lewis, M., Kane, R., and Haynes, R.M. (2008), 'An R package to compute commonality coefficients in the multiple regression case: An introduction to the package and a practical example', Behavior Research Methods, Vol. 40, No. 2, pp 457 – 466.

Nyblom, J. (1989). Testing for constancy of parameter over time, J of American Statistical Assoc., 84, 223-230.

Oberhofer, W. and Kmenta, J. (1974), "A general procedure for obtaining maximum likelihood estimates in generalized regression models", Econometrica 42, 579-590.

Pagan, A.R. and Wickens, M.R. (1989), 'A Survey of Some Recent Econometric Methods', Economic Journal, 99, pp. 962-1025.

Page, E.S., (1954). Continuos inspections schemes, Biometrika, 41, 100-114.

Page, E.S., (1955). A test for a change in parametric occurring at an unknown point, Biometrika, 42, 523-527.

Page, E.S., (1957). On problems on which a change in a parameter occurs at an unknown point, Biometrika, 44, 248-252

Palm, F.C. and Sneek, J.M. (1984), 'Significance Tests and Spurious Correlation in Regression Models with Autocorrelated Errors', Statistische Hefte, 25, pp. 87-105.

Parikh, K. S. (2000), India development report 2000, Oxford University Press, New Delhi.

Parikh, K. S., and Radhakrishna, R. (2002), India development report 2002, Oxford Universit Press, New Delhi.

Ploberger, W. (1989). The local power of the CUSUM-SQ test against hetroscedasticity, in Hackl, P. (edit) Statistical Analysis and Forecasting of Economic Structural Change, pp. 127-134.

Pfefferman, D. (1985), 'Regression Models for Grouped Population in Cross-Section Surveys', International Statistical Review, 53, pp. 37-59.

Prakash, S. (1979), 'Contributions to Bayesian Analysis in Heteroscedastic Models', unpublished Doctoral Dissertation (University of New England, Armidale, NSW, Australia).

Raj, B. (1989), "The peril of underestimation of standard errors in a random-coefficients model and the bootstrap", unpublished manuscript. Invited paper presented in the Department of Econometrics, Monash University, 1989

Rao, C.R. (1962), 'Efficient Estimates and Optimum Inference Procedures in Large Samples', *Journal of the Royal Statistical Society*, B, 24, pp. 46-72.

Rao, C.R. (1963), 'Criteria of Estimation in Large Samples', Sankhȳa, 25, pp. 189-206.

Rao, C.R. (1973), *Linear Statistical Inference and Its Applications.*, John Wiley, New York.

Rao, C.R., and Kleffe, J. (1980), 'Estimation of Variance Components', in *Handbook of Statistics*, Krishnakumar, P.R. (ed.), 1, pp. 1-40, North Holland.

Rao, J.N.K., Sutradhar, B.C. and Yue, D. (1993), 'Generalised Least Square F-Test in Regression Analysis with Two-Stage Cluster Samples', *Journal of the American Statistical Association,* vol. 88, no. 424 pp. 1388-1391.

Royall, R.M. (1968), 'An Old Approach to Finite Population Sampling Theory', *Journal of the American Statistical Association,* 63, pp 1269-1279.

Royall, R.M. (1970), 'On Finite Population Sampling Theory Under Certain Linear Regression Models', *Biometrika*, 57, pp. 377-387.

Royall, R.M. and Cumberland, W.G. (1981), 'An Empirical Study of the Ratio Estimator and Estimators of its Variance', *Journal of the American Statistical Association*, 76, pp. 66-87.

Sampson, A.R. (1976), 'Stepwise BAN Estimators for Exponential Families with Multivariate Normal Applications', *Journal of Multivariate Analysis*, 6, pp. 167-175.

Sampson, A.R. (1978), 'Simple BAN Estimators of Correlations for Certain Multivariate Normal Models with Known Variances', *Journal of the American Statistical Association,* 73, pp. 859-862.

Särndal, C.E. (1978), 'Design Based and Model Based Inference in Survey Sampling', *Scandinavian Journal of Statistics*, 5, pp. 27-52.

Schechtman, E. (1983). "A conservative nonparametric distributionfree confidence bound for the shift in the change point problem", *Comm Statist.* A 12,2455-2464

Scott, A.J. and Holt, D. (1982), 'The Effects of Two-Stage Sampling on Ordinary Least Squares Methods', *Journal of the American Statistical Association*, 77, pp. 848-854.

Scott, A. J. and Knott, M. (1974). "A cluster analysis method for grouping means in the analysis of variance", *Biometrika*, 30,507-512.

Scott, A.J. and Smith, T.M.F. (1969), 'Estimation in Multistage Surveys', *Journal of the American Statistical Association*, 64, pp. 830-840.

Searle, S.R. and Henderson, H.V. (1979), 'Dispersion Matrices for Variance Components Models', *Journal of the American Statistical Association*, 74, pp. 465-470.

Selliah, J.B. (1964), 'Estimation and Testing Problems in a Wishart Distribution', Technical Report No. 10, Stanford University.

SenGupta, A. (1981), 'Tests for Standardized Generalized Variances of Multivariate Normal Populations of Possibly Different Dimensions', Technical Report No. 50, Department of Statistics, Stanford University.

SenGupta, A. (1983), 'Generalized Canonical Variables', in Johnson, N.L. and Kotz, S. (eds.), *Encyclopedia of Statistical Sciences*, 3, pp. 326-330, Wiley, New York.

SenGupta, A. (1987), 'On Tests for Equicorrelation Coefficient of a Standard Symmetric multivariate clusterDistribution', *Australian Journal of Statistics*, 29, pp. 49-59.

SenGupta, A. (1988), 'On Loss of Power Under Additional Information -An Example', *Scandinavian Journal of Statistics*, 15, pp. 25-31.

SenGupta, A., and Vermeire, L. (1986), 'Locally Optimal Tests for Multiparameter Hypotheses', *Journal of the American Statistical Association*, 81, pp. 819-825.

Shao, J., Wang, H. (2002), 'Sample Correlation Coefficients Based on Survey Data under Regression Imputation', *Journal of the American Statistical Association*, Vol, 97, No. 458, pp 544 – 552.

Shively, T.S. (1988), 'An Exact Test for Stochastic Coefficient in a Time Series Regression Model', *Journal of Time Series Analysis*, 9, pp. 81-88.

Shively, T.S., Ansley, C.F. and Kohn, R. (1990), 'Fast Evaluation of the Distribution of the Durbin-Watson and Other Invariant Test Statistics in Time Series Regression', *Journal of the American Statistical Association*, 85, pp. 676-685.

Silvapulle, P. (1991), 'Point-Optimal Tests for Non-nested Time Series Models and Illustrations', unpublished Doctoral Dissertation, Monash University, Clayton.

Srivastava, M.S. (1965), 'Some Tests for the Intra-Class Correlation Model', *Annals of Mathematical Statistics*, 36, pp. 1802-1806.

Sukhatme, P.V. (1954), *Sampling Theory of Surveys, with Applications*, Ames, State College Press, Iowa.

Swamy, P.A.V.B. (1970), 'Efficient Inference in Random Coefficient Regression Models', *Econometrica*, 38. pp. 311-323.

Swamy, P.A.V.B. and Arora, S.S. (1972), 'The Exact Finite Sample Properties of the Estimators of Coefficients in the Error Components Regression Models', *Econometrica*, 40, pp. 253-260.

Tauchen, G. (1985), 'Diagnostic Testing and Evaluation of Maximum Likelihood Models', *Journal of Econometrics*, 30, pp. 415-443.

Votaw, D.F. (1948), 'Testing Compound Symmetry in a Normal Multivariate Distribution', *Annals of Mathematical Statistics*, 19, pp. 447-473.

Wallace, T.D. and Hussain, A. (1969), 'The Use of Error Components Models in Combining Cross Section with Time Series Data', *Econometrica*, 37, pp. 55-72.

Walsh, J.E. (1947), 'Concerning the Effect of Intra-Class Correlation on Certain Significance Tests', *Annals of Mathematical Statistics*, 18, pp. 88-96.

White, H. (1982), 'Maximum Likelihood Estimation of Misspecified Models', *Econometrica*, 50, pp. 1-25.

Wilks, S.S. (1946), 'Sample Criteria for Testing Equality of Means, Equality of Variances, and Equality of Covariances in a Normal Multivariate Distribution', *Annals of Mathematical Statistics*, 17, pp. 257-281.

Williams, P. and Sams, D. (1981), 'Household Headship in Australia: Further Developments to the IMPACT Projects' Econometric Model of Household Headship', IMPACT Project Research Centre Working Paper BP-26, University of Melbourne.

Williams, E.J. and Yip, P. (1989), 'Conditional Inference About an Equicorrelation Coefficient', *Australian Journal of Statistics*, 31, pp.138-142.

Worsley, K. J. (1983). "The power of likelihood ratio and comulative sum tests for a change in a binomial brobability", *Biometrika*, 70, 455-464.

Worsley, K. J. (1986). "Confidence regions and tests for change point in a sequence of exponential family random variables", *Biometrika*, 73, 91-104.

Wu, P.X. (1991), 'One-sided and Partially One-sided Multiparameter Hypothesis Testing in Econometrics', unpublished Doctoral Dissertations, Monash University, Clayton.

Wu, P.X. and Bhatti, M.I. (1994), 'Testing for Block Effects and Misspecification in Regression Models based on Survey Data', *Journal of Statistical Computation and Simulation*, Vol.50, Nos. 1-2, pp.75-90.

Wu, C.F.J., Holt, D. and Holmes, D.J. (1988), 'The Effect of Two-Stage Sampling on the F Statistics', *Journal of the American Statistical Association*, 83, pp. 150-159.

Yates, F. (1949), *Sampling Methods for Census and Surveys*, 1st Edition, 2nd Edition (1953); 3rd Edition (1969), Griffin, London.

Zacks, S. (1983). Survey of classical and Bayesian approach to the change point problem: Fixed sample and sequential procedures of testing and estimation, *Recent Advances in Statistics*, Chapter in honor of Herman Chernoff's Sixtieth Birthday (Rizvi M. H. ed.) Academic Press, New York, 245-269.

Zacks, S. (1991). Detection and chang-point problem, Handbook of sequential analysis (Ghosh, B.K. and Sen, P.K. eds.) Series Statistics, vol. 118, M. Dekker, New York, 531-562.

INDEX